Forces Shaping the U.S. Academic Engineering Research Enterprise

Committee on Forces Shaping the
U.S. Academic Engineering Research Enterprise

National Academy of Engineering

NATIONAL ACADEMY PRESS
Washington, D.C. 1995

NATIONAL ACADEMY PRESS • 2101 Constitution Avenue, NW • Washington, DC 20418

NOTICE: The National Academy of Engineering was established in 1964, under the charter of the National Academy of Sciences, as a parallel organization of outstanding engineers. It is autonomous in its administration and in the selection of its members, sharing with the National Academy of Sciences the responsibility for advising the federal government. The National Academy of Engineering also sponsors engineering programs aimed at meeting national needs, encourages education and research, and recognizes the superior achievements of engineers. Dr. Robert M. White is president of the National Academy of Engineering.

This publication has been reviewed by a group other than the authors according to procedures approved by a National Academy of Engineering report review process.

Partial funding for this effort was provided by the National Science Foundation.

Library of Congress Catalog Card Number 95-69121
International Standard Book Number 0-309-05284-X

Printed in the United States of America

Preface

On February 18, 19, and 20, 1994, the National Academy of Engineering (NAE), with funding from the National Science Foundation (NSF), convened a public symposium and workshop on the forces shaping academic engineering research in the early 1990s and beyond. The report that follows has been prepared by an NAE committee charged with organizing the symposium and workshop and reporting back to the NSF. The membership of the Committee on Forces Shaping the Academic Engineering Research Enterprise is listed on page v of this volume.

In preparing this report, the committee drew heavily on the symposium presentations and workshop discussions. Nonetheless, the committee is the author of this report and is responsible for its arguments and findings. The papers presented at the public symposium as well as a background paper prepared for workshop participants follow the committee's report.

It is important to note that this document makes no claim to be an exhaustive examination of the issues facing academic engineering research. For example, there is no focus in this report on the impact of changing demographics on engineering students or faculty, or on the effect of the military build down on the character of the national portfolio of engineering research. The intent was not to be comprehensive, and the committee was not asked or constituted to write the last word on the status and future of academic engineering research.

On behalf of the National Academy of Engineering, I would like to thank the authors of the papers and the chairman and the members of the committee for their insights and efforts on this project. In addition, I would like to thank Bruce Guile, Debbie Stine, and Jessica Blake for their excellent staff work on this project.

Robert M. White
President
National Academy of Engineering

COMMITTEE ON FORCES SHAPING THE U.S. ACADEMIC ENGINEERING RESEARCH ENTERPRISE

WILLIAM R. SCHOWALTER, *Chairman*, Dean, College of Engineering, University of Illinois
DANIEL C. DRUCKER, Graduate Research Professor of Engineering Sciences, University of Florida
ALEXANDER H. FLAX, Senior Fellow, National Academy of Engineering
C. WILLIAM GEAR, President, NEC Research Institute, Inc.
PAUL C. JENNINGS, Vice President and Provost, California Institute of Technology
SIMON OSTRACH, *NAE Home Secretary*, Wilbert J. Austin Distinguished Professor of Engineering, Case Western Reserve University
A. RICHARD SEEBASS, Professor of Aerospace Engineering Sciences, University of Colorado
JOHN A. WHITE, JR., Dean, College of Engineering, Georgia Institute of Technology

NAE Staff

BRUCE GUILE, Director, Program Office
DEBORAH STINE, Project Director
JESSICA BLAKE, Project Assistant

Contents

Introduction

The way in which academic engineering research is financed is changing at an unprecedented rate. So, too, are public expectations for the outcomes of such research. One can relate these changes to the overlap of two unrelated occurrences: the end of the Cold War—expected to cause a drop in support for defense-related research in universities and an immediate loss of appetite for highly trained engineers in the defense industry; and a realization in corporate America, over a more extended time period, that many major U.S. producers of technological products were not competitive in a global economy. Both of these factors have affected greatly the nation's economy. Nowhere are these effects taken more seriously than in the advanced education of engineers and scientists, and not without good reason. For example, while senior officials at the Department of Defense (DOD) have declared their intention to maintain DOD support for university basic research in spite of large reductions in the overall defense budget (Adams, this volume), the U.S. House of Representatives cut almost $1 billion in DOD funds for university research in its version of the 1995 appropriations bill. Fortunately, most of this money was restored by House Senate conference committee.

Because research universities are neither listed on stock exchanges nor subjected to the scrutiny of financial analysts, the general public and policymakers have been largely unaware of the shock wave of apprehension currently traveling through academia. Although the full impact of these changes has not yet been felt, there are some significant indicators of problems. For instance, while the award of masters and Ph.D. degrees to U.S. citizens has been increasing, the annual number of U.S. undergraduate engineering graduates has dropped by 15 percent since its peak in 1985.

The anticipated shift away from the support of academic research is by no means limited to engineering; it extends to many fields of scientific research. However, academic engineering differs from academic science both in its intrinsic ties to socioeconomic goals and in the mechanisms and time scales by which it can respond to changes in these goals. Indeed, most engineering research is closer to application in both time and concept. The changes in attitude and policy toward academic engineering research provide an opportunity for those involved in the enterprise to reinvent its mission and reevaluate its activities. In this report, the committee takes these external changes as givens and suggests ways in which the products of academic engineering research and education can be designed to be consistent with the long-term health of the nation. The report makes several recommendations to ensure there is sufficient appropriately trained technical talent to meet national social and economic goals, to maintain a position of U.S. leadership in the global economy, and to preserve and enhance the nation's engineering knowledge base:

- Some important stakeholders, including industry and government, have abandoned or reduced their stewardship of fundamental engineering research. Others appear to be retreating from their long-term commitments to the continuing viability of academic research. This could put at risk the nation's primary means for attracting talented minds to professional careers at the leading edge of technology development. The result may be a failure to maintain the knowledge base on which technological supremacy rests. Neither can be allowed to happen if the United States is to retain its technological competitiveness.
- There is an intimate relationship between academic engineering research, the quality of engineering graduate education, the nation's industrial infrastructure, and economic growth. Therefore, it is critical that universities examine their processes for producing Ph.D.'s. Academic institutions need to determine whether the research portfolio and related instructional practices of engineering faculty are contributing adequately to the education of graduate students. Specifically, do these students have the skills, knowledge, and, most important, the orientation to be of direct value to potential employers in both the near and the long term?
- In view of the economic value of close, effective university-industry research relationships for both education and development of the nation's engineering knowledge base, it is critical that universities and companies commit themselves to bold new efforts at collaboration. Under the leadership of the National Science Foundation (NSF), Engineering Research Centers have stimulated the development of government-industry research link-

ages (Lane, this volume). According to a recent study (Dickens, this volume), there are 281 university research centers sponsored by six federal agencies (including NSF) and over 1,000 university-based engineering research units in the United States. Most of these research units were established as university initiatives in the past 10 years, and their success in establishing industry linkages varies widely. Much broader adoption of such linkages by industry—without government sponsorship and participation—is needed.

- Consistent with the important role of academic engineering research in the advancement and diffusion of the engineering knowledge base and the training of engineers, substantial increases are needed in the level of support for academic engineering research and associated aspects of engineering education. Such increases will enhance U.S. leadership in commercially important technologies, improve industrial competitiveness, and increase economic growth. Reports issued over the past decade by the National Academy of Engineering, the National Research Council Engineering Research Board, and the National Science Board Committee on Industrial Support for R&D all have echoed the need to boost funding in this area (Committee to Evaluate the Programs of the National Science Foundation Directorate for Engineering, 1985; National Research Council, 1987; National Science Board, 1992).
- Because policymakers tend to be unaware of the variety of purposes and products of government-sponsored research, the engineering community must coordinate and focus more effectively the many voices speaking for engineering. Both policymakers and the public need to better appreciate the important differences between scientific and engineering research, especially with regard to how quickly the two disciplines can address pressing national concerns.

In general, the concept of engineering research is not readily understood. In academic settings, its distinction from research in the basic sciences is even less well understood. Therefore, the next section of this report is devoted to an exposition of the nature and value of academic engineering research.

WHAT IS ENGINEERING RESEARCH AND HOW DO ENGINEERING AND SCIENCE INTERACT?

In many ways, the methods of academic engineering research and the resulting insights into the nature of the physical world are indistinguishable from those of basic scientific research. However, there are crucial differences between the two endeavors. Basic scientific research is concerned with the discovery of new phenomena and their integration into coherent conceptual models of major physical or biological systems. By definition, the focus of greatest interest tends to be at the outer edges of present knowledge. Most scientific knowledge will, in a highly variable and unpredictable fashion, find technical applications of economic and social value, but in most cases the nature of such applications will not be apparent to the those who perform the original scientific research.

Basic research in engineering is by definition concerned with the discovery and systematic conceptual structuring of knowledge. Engineers develop, design, produce or construct, and operate devices, structures, machines, and systems of economic and societal value. Virtually all engineering research is driven by the anticipated value of an application. However, not all potential applications can be anticipated, and occasionally the hoped-for application may not be nearly as important as one that turns up by serendipity. The time from research to production may be a few years, as in the development and application of the laser or in the progression from the integrated circuit to microprocessor, or it may be decades, as in the development of television.

Engineering, unlike science, is concerned not only with knowledge of natural phenomena, but also with how knowledge can serve humankind's needs and wants. Such variables as cost, user compatibility, producibility, safety, and adaptability to various external operating conditions and environments must be taken into account in the design, development, operational support, and maintenance of the products and services that engineers create. Thus, engineering involves the integration of knowledge, techniques, methods, and experiences from many fields.

Also, almost all university research in both science and engineering is performed as a component of the advanced education of students. For most engineering students, the goal of a career in industry motivates their pursuit of advanced study, and this will increasingly be the case in the future. Because of this, engineering students' outlook on research tends to be predisposed toward application in engineering practice.

Basic science and mathematics have advanced rapidly in the past several decades with the development of computers that can deal with increasingly complex problems. At the same time, engineering science, research, and practice have employed increasingly advanced analytical and experimental methods across the spectrum of engineering fields and industrial sectors. In *What Engineers Know and How They Know It* (Johns Hopkins University Press, 1990), Walter Vincenti has identified some theoretical and experimental features common to both scientific and engineering research. In fact, in some engineering fields such as electronic materials, the analytical and experimental methods and instruments used may be indistinguishable from those in the basic-science fields of solid-state physics and chemistry.

WHAT HAS ENGINEERING RESEARCH DONE?

Engineering education and academic engineering research have played important roles in shaping this nation's industrial capabilities. They are doing so to an increasing degree as more technically advanced and complex products and systems are emerging in the marketplace and in the social and economic infrastructure. As new knowledge and more powerful analytical and experimental methods expand the power of engineering in practice, problems of design and development once considered too complex to be dealt with other than empirically, intuitively, or by trial and error have become solvable.

As Simon Ostrach points out (this volume), in many instances, industry lagged in its awareness of this new problem-solving capacity and in its readiness to adopt new methods. Engineers engaged in academic research, industrial research, or product and system design, development, and innovation were needed to assemble, evaluate, and exploit the full range of available scientific and engineering knowledge and methods in their work. This was true whether their work was directed toward the near or long term.

In a number of cases, at relatively long intervals and usually at a relatively slow rate, entirely new technologies leading to new products and services have emerged from basic scientific research. Thus, the development of modern broadcast radio and TV evolved over many decades from the early work of Maxwell and Hertz in the nineteenth century. To achieve economic and societal utility from these elements of fundamental scientific knowledge required research interspersed with inventions relating to circuit design, amplifiers, vacuum tubes, feedback and circuit stability, antennae, and amplitude and frequency modulation, among other things. Edison, Marconi, DeForrest, Armstrong, Fessenden, Nyquist, and Bode all contributed to the variety of achievements that led ultimately to the modern attributes of broadcast radio. Their basic research and invention were clearly aimed at achieving applications in communications technology and come under the mantle of engineering rather than science. However, there is a close coupling between scientific and engineering research. Refinements in the quality and performance of such things as microwave tubes and devices, electronic instrumentation, and computers, which come out of engineering, nourish the progress of scientific research. The resulting new scientific principles can in turn facilitate engineering research and development on new processes, devices, and instruments.

Knowledge derived from research does not necessarily or uniformly flow from science to engineering. Engineering progress based on empirical, experimental, and heuristic methods often anticipates underlying scientific principles. Thus, the development of the airplane by the Wright brothers preceded fundamental aerodynamic theories and principles adequate for the design of either airplane wings or propellers. Nevertheless, engineering development techniques, including the use of wind tunnels and flight tests (of gliders), enabled the Wright brothers to design a flyable, controllable machine. Subsequent research, largely in engineering but also in some of the basic sciences, has made possible the tremendous growth in global air transportation over the past century. Engineering research aimed at achieving technical and economic progress of this sort must go well beyond the limited knowledge on which invention or demonstration of technical feasibility of a new device, machine, or system is based. It must produce more in-depth and usually more quantitative information that will allow for continuing improvements in the performance, economics, and range of application of the original invention or technical demonstration. Progress in the development of prime movers and power plants—from steam engines to internal combustion engines and gas turbines—was mainly the result of engineering research and development, although advances in engine and turbine materials benefited from scientific research in physics and chemistry. Recent advances in high-strength, high-stiffness fiber composite materials flowed initially from engineering research.

The development of practical electronic computers was also aided by engineering research, along with mathematics (programming concepts and software development) and solid-state physics (transistors). The most significant recent advances in computers have followed from the development of integrated circuits and microprocessors, both products of engineering research. The sequence was: transistor, 1948; integrated circuit, 1959; microprocessor, 1972. Transistors, integrated circuits, and microprocessors have not only had a profound influence on computers but, through engineering application as components, have also brought about major advances in a broad spectrum of products and services, from telecommunications to transportation and industrial manufacturing and process control.

Computers themselves, of course, have affected the course of scientific research in fields as diverse as astronomy and solid-state physics. The work that led to the invention of the electronic computer was university based. On the other hand, the invention of the integrated circuit took place in industry. In both cases, their subsequent development and widespread application in industrial products and infrastructure owe much to the emergence and diffusion of systematic, rationally based methods of analysis and design for both hardware and software. University research and education played indispensable roles in this process.

Armstrong (this volume) points out that university-based hardware research no longer is the major contributor to computer development that it was in the early days of the computer industry. This is to some degree typical of new technologies that originate mainly from university research and then mature in industry. A similar scenario has played out in the fields of artificial intelligence, neural networks, and several other advancements in computer architecture and software. Some recent developments such as RISC were the result of university-industry collaboration (Tien, this volume).

Universities continue to play a role in the systematic organization, extension, and explication of engineering knowledge. Through the involvement of graduates and faculty, and via the influence of published research, universities will remain important in many industrial sectors long after early-stage academic research has found its way into an industrial product. Armstrong also cites the general utility of graduate education in scientific and engineering fields, which goes beyond the specific technical content included. It imparts to graduates ways of approaching and solving problems using powerful and fundamental principles. These attributes qualify Ph.D.'s for many positions in the socioeconomic system outside of traditional R&D. With the growing importance of technology in every field of human activity, the opportunities for engineering Ph.D.'s in these nontraditional positions will grow, even if the number of traditional R&D positions declines.

To summarize, the value of engineering research is its capacity to solve real-world problems. Engineering research has provided the systematic underpinnings for the design, analysis, production, and operation of products and systems. Academic engineering research has been academic only in its setting and time frame; first-rank academic engineering research is focused by goals of synthesis, design, analysis, production, and operation but may be too risky, too hard, too general, or too far ahead in time from market application to interest engineering researchers working for private industry. Also, academic engineering research provides the setting for advanced training and education of our nation's most able technical specialists. It is from this reservoir of talent that the most creative technical ideas which underpin industrial progress and economic growth have emerged.

WHY IS ACADEMIC ENGINEERING RESEARCH AT RISK, AND WHY SHOULD ITS HEALTH BE PRESERVED?

Academic engineering research has been funded primarily by the federal government. All research universities have benefited from the support of industry, and in some instances, states have funded projects aimed at transportation, environmental concerns, or other local issues. But it is federal agencies, often branches of the Department of Defense, that contribute the largest share—57 percent in 1992[1]—of the total spent on academic engineering research. (See Dickens' Table 5 this volume for closely similar data for 219 engineering research universities.) The federal government also pays a considerable portion of the support for graduate students' education.

Most graduates of Ph.D. programs in engineering enter the industrial sector upon completion of their studies. However, because their support derives mainly from government funding and the subject of their research is the result of a compact between their faculty adviser and a government agency, the stake of industry has been indirect. On the one hand, industry receives a government subsidy in the form of educational and research support provided to the graduate. Graduate research itself often adds significantly to the fundamental knowledge base, enabling industry to extend its own research and development. On the other hand, industry often has little influence on the direction taken by academic research, and university-trained students often have no appreciation of the constraints and drivers affecting the conduct of research by industry, or indeed of why industry should even have a stake in research. Simply put, there has been in many fields a fundamental disconnect between industry's needs and government's support for academic engineering research. This is by no means the state of affairs for *all* academic engineering research. Nevertheless, the picture described above has been, at the very least, not unusual in some of the nation's most renowned research universities.

These issues could be tolerated in an era when federal and corporate budgets were ample and engineers had a wide choice of jobs in defense- or civilian-related industries. But the pressure of global competition and, more recently, the threat of major reductions in defense-related R&D funding have driven much of corporate America into a survival mode. One of the segments of corporate activity most vulnerable to such pressures is research. Lofty expressions of the need for a corporation to invest in its future, to nurture long-range thinking, and to hire the best minds of today's new engineering talent pool will rarely, in a boardroom discussion, hold sway over the requirement to keep the company solvent for the next quarter.

The result, still hidden from much of the general public and policymakers but becoming painfully clear to the best and brightest of America's entry-level advanced-degree engineers, is a phenomenon resembling the pileup at the end of a down escalator when those emerging from advanced engineering education do not keep moving into the corporate world. The pileup should be temporary, however, as the forces of a free-market economy cause those intellectually qualified to ride the escalator to turn to other pursuits. There are three aspects of this natural consequence of supply and demand that are unsettling: the time required to adapt to future increased demand (engineers require more than 5 years, on average, to obtain a Ph.D. after earning a bachelor's degree); reduced creative contributions to the nation's welfare; and an impending dearth of contributions to the fundamental engineering knowledge base currently fed, in large measure, by the research conducted in universities by graduate students.

Part of the rationale for the post–World War II compact between Congress, government agencies, and universities was that it maintained the "engine of knowledge creation." The output of the engine was not only new knowledge available to all through scholarly literature and technical meetings, but also knowledgeable persons ready to enter the spectrum of technological endeavor with a proven capacity for formulating and solving complex technical problems at the limits of the existing state of human understanding.

No nation expecting to use the world's storehouse of fundamental knowledge to its competitive advantage can, over the long term, afford not to contribute to that storehouse. In particular, nations at the forefront of technological development are always in the position of being caught by other nations and so must aggressively exploit technological advances to stay ahead. Moreover, if the storehouse of fundamental knowledge is not being resupplied, the young minds best able to contribute to engineering creativity will never be attracted to engineering research in the first place. Clearly, it is in the nation's interest to preserve a reasonable pipeline of knowledge and intellect of the types described above. Because of the forces disrupting the traditional ways of supporting such a pipeline, the academic engineering research community and the government and private clients dependent upon this community are asking the obvious questions: What is a "reasonable" pipeline? and Who should pay for its maintenance?

In the diverse system of education that exists in the United States, quantitative answers to these important questions neither can be nor should be determined by a committee. However, a knowledgeable committee may suggest approaches for offsetting some of the negative factors and trends now being faced by the academic engineering community.

The role of academic research is multifaceted. It serves to expand the engineering knowledge base; contributes to the exploration and application of specific areas of technology; provides systematic contexts and infrastructure for the diffusion and transfer of engineering and technological information; and provides training for most of the future leaders in engineering across the spectrum of research, development, design, and other engineering functions. Because of this varied role, leading engineers have long believed that academic engineering research (and in many instances engineering research in general) is underappreciated and undersupported. Over the past decade, several groups of leading U.S. engineers have recommended that the level of funding for engineering research be substantially increased in the interests of U.S. technological leadership, international competitiveness, and economic growth. These recommendations have been put forward by the Committee to Evaluate the Programs of the National Science Foundation Directorate for Engineering, the National Research Council Engineering Research Board, and the National Science Board Committee on Industrial Support for R&D, among others (see References). Although often accepted in principle by government, university, and industry sponsors of research, fiscal exigencies in each of these sectors have tended to limit the implementation of these recommendations. This committee, too, believes it is a matter of high national priority to enhance funding for engineering research.

RESPONDING TO THE CHANGED ENVIRONMENT FOR ACADEMIC ENGINEERING RESEARCH

Many of the current pressures on universities relate to the management of the university enterprise, the strength and character of ties to the industrial community, and the character of undergraduate education. With regard to academic engineering research, the touchstone issue is the education-research nexus in graduate engineering education and its relation to industry needs in the global economy. Does the focus on academic research contribute effectively to graduate engineering education? What are the strengths and weaknesses of current academic engineering research programs? How do we overcome the disconnect between academic engineering research and its relevance to careers in industry? Given the changing nature and potential lessening of federal support for fundamental engineering research, is it realistic to expect industry to exert more control over such research by investing more in its support? If industry does this, how can the long-term interests of the nation best be served?

Research and Graduate Education in Engineering

In evaluating the contributions of academic engineering research to national goals, a major question is the degree to which such research helps those individuals who will, whether they join academia, industry, or government, enhance and apply the knowledge base relevant (in either the long or short term) to the technical problems facing the country. In recent years, the academic research enterprise has often been judged by engineers and managers outside academia to be too narrow and detached from application and practice. The concern over the relevance of academic engineering research is especially strong because the enterprise is being asked both to move closer to short-term problems and to justify itself mainly on the basis of near-term contributions.

There was considerable agreement among workshop participants that the nation's engineering schools should be challenged to construct their curricula in ways that address real-world problems. One workshop breakout group proposed three guiding principles for engineering faculty:

- Connect practice to teaching;
- Connect research both to current and likely future real-world problems; and
- Connect teaching to research.

Although simply expressed the concept is sound: The character of engineering and engineering research demands that real problems be important elements in academic research and the training of engineers. This trilogy of precepts should not be interpreted to imply that all engineering research, much less all engineering knowledge of practice, should flow entirely from within universities or from university faculties. The consensus of the workshop was that stronger and better interactions with industry are essential. (This point is not unique to the field of engineering; medical research, education, and practice are somewhat similar.) Here again, we come to the pervasive influence of *motivation* that distinguishes academic engineering from academic science and instills in advanced engineering students a mindset and problem-solving methodology oriented to the creative innovation process for new products and services—a mindset noticeably different from that of their counterparts in science.

Therefore, the committee recommends that engineering schools examine their processes for producing Ph.D.'s and determine whether the research portfolio and related instructional practices of the engineering faculty are contributing adequately to graduate students' skills and knowledge. Most important, academic institutions must turn out engineers who are of value to potential employers in both the near and long term.

The hallmark of well-educated graduate engineers should not be restricted to narrow technical or research competence. More important is the ability to solve tough technical problems requiring the application of a broad range of research skills and experience along with an understanding and appreciation of design, development, and creative innovation, which contribute to the solutions of these problems. Do graduates have adequate focus-knowledge of process? Are they narrow specialists able primarily to make contributions to the development of the discipline, or are they broadly educated problem solvers who have acquired deep research competence through their graduate education? What can be done by the university and the faculty to ensure that the research agendas of the engineering faculty are in concert with the needs of the educational process for Ph.D.'s who will join industry, government, or academia?

New and Newly Formulated University-Industry Collaborative Activities

Engineering research serves private for-profit institutions as well as the public interest in such areas as economic development, environmental protection, national defense, and health care delivery. In engineering, perhaps more than any other technical discipline in a university, there can be great value in joint university-industry activities on a long-term basis. Indeed, historically, engineering colleges and industries have evolved together as technology changed. Collaboration between industry and academia in engineering research, education, and practice has been rich and varied. Now, as national challenges shift with the end of the Cold War and the globalization of industry, the relative responsibilities of industry, academia, and government for the engineering knowledge base must also shift. That shift (or steady evolution) will be worked out in thousands of new and altered relationships among academic engineering researchers, industry, and government. What direction should such changes take? If left entirely to the motivations of individual researchers, universities, companies, and government agencies, will the system evolve in a direction that can serve the national interest effectively? What should be the character of university-industry relationships, and what should be the role of government in academic engineering research?

Weak links between industry and universities in areas in which industry could benefit from the knowledge generated by academic engineering research are evidence of failure on the part of both sectors. No single model of collaboration will work for every situation. What must be recognized is that industry and universities, working together, need to evolve bold new joint ventures to fill the void left by withdrawal of a mainly government-fueled system. New models of collaboration cannot be spelled out in detail here; however, it is clear that among their necessary conditions are long-term commitment, personnel exchange, and recognition that academic engineering research is a component of the education of the nation's technical talent pool for the next generation. Tien (this volume) calls attention to the need for academic researchers not only to carry out work in collaboration with industry, but also to take part in "missionary work to convince people to use the ideas."

Therefore, the committee recommends that universities and companies commit themselves to relationships that couple industrial technology and practices with the leading edge of research and advanced education in engineering.

Workshop participants recognized that the transition to a new (or perhaps previous historical) relationship between academia and industry is not something that can occur in the absence of substantial change in the structure of institutions. It is clear that tenure and promotion policies in universities need to be reassessed with respect to the degree to which they give balanced weight to research, teaching, and linkages to industry and public policy. There is a need for industrial commitments that transcend but do not undercut short-term competitive advantage. That is done today in many industry sectors, such as safety and the environment. No one questions the importance of these issues to the nation's social well-being or to a company's long-term self-interest. The same considerations apply to maintaining the quality of the technical talent pool to ensure the nation's industrial and economic health.

Conclusion

Academic engineering research is at a point of evolutionary change that will determine its character and its intensity for the next several decades.

In contemporary business terms, changes in its customer base have left the enterprise without a sufficient set of stakeholders. It is time for those who wish to be stakeholders to come forward and to claim ownership. Because the long-term technical health of the nation is clearly dependent on the future health and direction of academic engineering research, the government must continue to be an important stakeholder, albeit with a role different from the one practiced during the last half-century. The time has arrived for government, industry, and universities to make collective and conscious decisions tailored to the strengths and needs of today's technology-dependent society and economy.

Finally, the workshop discussion confirmed the complexity of articulating the difference between academic engineering research and its scientific counterpart. Participants became firmly convinced of the need to explain these differences to those responsible for policies that affect both sectors.

Therefore, the committee recommends that the engineering community coordinate and focus more effectively the many voices speaking for engineering. Government leaders in Washington and the public at large must understand the important differences between scientific and engineering research. Only then will the special character of engineering research and education in meeting the needs of the national industrial economy, societal infrastructure, and public health and safety be fully appreciated.

NOTE

1. The average share of academic engineering research supported by federal agencies is 57 percent, and the spread is quite wide: 43 percent of civil engineering and 76 percent of aeronautical engineering research in academia is supported by federal agencies. National Science Foundation. 1992. Selected Data on Academic Science and Engineering R&D Expenditures: 1992 (Tables 5 and 6). Surveys of Science Resources Series. Washington, D.C.

REFERENCES

Committee to Evaluate the Programs of the National Science Foundation Directorate for Engineering. 1985. New Directions for Engineering in the National Science Foundation. Washington, D.C.: National Academy of Engineering.

National Research Council. 1987. Directions in Engineering Research. Commission on Engineering and Technical Systems, Engineering Research Board. Washington, D.C.: National Academy Press.

National Science Board. 1992. The Competitive Strength of U.S. Industrial Sciences and Technology: Strategic Issues. Committee on Industrial Support for R&D. Washington, D.C.: National Science Foundation.

Academic Engineering Research in a Changing World

Neal F. Lane

The Office of Science and Technology Policy convened a Forum on "Science in the National Interest: World Leadership in Basic Science, Mathematics, and Engineering," which was in Washington at the end of January. More than 200 scientists, engineers, educators, administrators, and science policy thinkers met at the National Academies in Washington and for two days discussed a broad range of issues facing the research community.

What I took away from the Forum was not a set of crisp insights into the optimal role of the federal government in supporting the research enterprise. That was not expected. Rather, I came away from the two days of discussions with an appreciation that there is a growing consensus on a number of issues, indeed challenges, that we in the research and education community must face. The public appears not to be convinced that incremental support for basic science, mathematics, and engineering should be a high priority in competition with programs that address immediate societal needs. Equally as important, however, is the consensus that we are doing many things well. So, when we look at ways that we might better respond to society's challenge, we must ensure that we protect and nurture the things that are working well.

Let me begin by noting some of the things we do well—and by "we" I am referring broadly to the science, mathematics, and engineering research and education community.

We do an exceptionally good job of educating graduate students to carry on the academic and other professional basic research tradition. We have an exemplary system of evaluating and funding quality research across the range of research topics through merit review using peer evaluation. We have developed a number of ways to capitalize on the intersection of interests of academic researchers, the industrial sector, and the national welfare.

These are some of the things, let's call them core values, we have developed over the years that must be preserved. One could sum them up by saying that our success in supporting research and graduate education reflects a commitment to excellence in the pursuit of new knowledge through exploration and discovery. Central to maintaining this commitment to excellence is our reliance on merit review and investigator initiated proposals. And as I said at the outset, excellence remains the key building block for the future, just as it has served the nation so well in the past.

Some would say that success on these criteria is more than sufficient. Or put in the words that are often invoked by those who would leave well enough alone. If it ain't broke, don't fix it.

While I would not contend that the research enterprise is somehow "broke," if it were to break, the future of this nation would be at great risk. So I think it is time we engaged in some thoughtful reconsideration, perhaps some preventive maintenance, and a rethinking of our role as researchers and educators in the larger society. The world in which our research enterprise has been so successful has changed. The boundary conditions are different, and it would be folly to ignore that.

I say this for the same reasons that were expressed at the Forum—considerations that have been voiced by a growing number of observers (and participants) in recent years. Of course at the top of most people's list of world changes that would cause us to rethink our purpose is the end of the Cold War. This change in the basis for much of our foreign and domestic policy over the past 45 years has resulted in a shift in our national outlook.

By extension, it has resulted in a continuing redistribution of our research resources, and a reconsideration of national security as a key rationale for supporting research. Congressman George Brown summed this up at the Forum by saying: "We cut our teeth as scientists on national security. Our job here today [meaning at the Forum] is to refocus that lens on a vastly different era for science for America." Mr. Brown kindly did not observe that we scientists and engineers are now "longer in the tooth" and so is the nation. Things do change!

A second category of reasons to reconsider science and technology policy is related to the first. The issue has taken on much more urgency at a time of consistent federal budget deficits. The public and its representatives in Congress have asked: What are the guiding principles behind the myriad federally supported research efforts? And what are we getting for our investments?

For those of us who are immersed in research, the answers to such questions seem obvious—after all, look around—we are getting space-age materials, fiber optics communication, supercomputing, advanced electronics, biotechnology, and insight into the makeup of the universe and ourselves. We are getting information superhighways, new manufacturing technologies, and an understanding about global change, including the impact of human behavior on the environment.

To the research community, these answers seem so obvious that we sometimes have had trouble believing the questions are sincere. But the questions are being asked by serious people and in a serious tone, although perhaps in a somewhat more subtle manner.

Over the past year or so, questions about how priorities are set within the research community and by the various federal agencies have come from a number of quarters in Congress. Various members have stressed their desire to have a research enterprise that is more reflective of and responsive to national societal goals.

One visible response to the need for more focus in research policy is evident in congressional willingness to look with favor on such activities as the former Federal Coordinating Council on Science, Engineering, and Technology (FCCSET, for short), which has since been superseded by the President's National Science and Technology Council (NSTC). The FCCSET process identified a number of areas for improved coordination and increased R&D investment—advanced manufacturing technology, global climate change, science and mathematics education, high-performance computing and communication, biotechnology, and advanced materials processing.

Many in the research community have expressed concerns that too much emphasis in these areas, particularly by the National Science Foundation (NSF), might shift funds away from some important disciplines or skew the type of research that would be funded. Concerns expressed in Congress heightened the anxiety, and were viewed by some as a harbinger of change for NSF and its role in support of research.

At the Forum, Senator Mikulski, who chairs the VA-HUD-Independent Agencies Appropriations Subcommittee, offered some clarification on the intent of her committee's report language. She emphasized that research in strategic areas is not meant to change NSF's role in supporting basic or fundamental research across the spectrum of science, mathematics, and engineering.

Rather, strategic areas provide a focus for the research and an impetus for viewing it in a larger perspective. This perspective includes the research community's obligation to consider its overall role in terms of how the fruits of research can be used to improve people's lives. Senator Rockefeller spoke at the Forum of the need to strengthen mechanisms for industry-university collaboration.

For the engineering research community, this is not much of a leap. At NSF we have a long history of successful university-based research centers pioneered by our Engineering Division. In the coming fiscal year we expect to have 22 Engineering Research Centers working with more than 600 industrial partners. Our Engineering Division has provided models for many successful partnerships between industry, universities, and the states, models that can lead the way for partnerships in other disciplines.

Although the "s" word as I call it, *strategic*, dominated many of our discussions, it was not the only item on the agenda, nor was it the only issue on which the participants felt they made progress.

Another important area of discussion was the development of human resources—particularly making the research enterprise more accessible and welcoming to women, underrepresented minorities, and persons with disabilities. Many at the Forum noted that opening up the research arena is not a choice between excellence and diversity, but a recognition that true excellence cannot exist without diversity.

A related concern is in improving science, mathematics, technology, and engineering education. NSF has been a leader in seeking ways to broaden the base of students who are interested and engaged in science topics. It is widely recognized that in an increasingly complex world, every person will benefit from a better understanding of basic mathematics and science concepts as the complexity of jobs increases.

Within the engineering community, NSF has supported new ways to approach the engineering curriculum and to expand professional opportunities in ways that reflect the diversity of our country. Joe Bordogna and his colleagues in the engineering directorate at NSF have done an outstanding job in this regard. The Engineering Education and Centers Division has established four—soon to be eight—innovative Engineering Education Coalitions to look at new ways to structure the engineering curriculum. These projects provide a much more integrated vision of engineering education, a vision that reaches out beyond the traditional pool of candidates for engineering and technology education.

True progress will require a cultural change—a change that internalizes within the community a commitment to encouraging more women and underrepresented minorities to view engineering as an appropriate, even desirable, career choice.

We should also seek to provide broader education and training to talented people who pursue graduate education but do not seek to fill academic positions after graduation. Highly educated and qualified students should have a variety of options open to them upon completion of their degrees.

Yet unfortunately, graduate education in many fields of science and engineering is cause for great concern. There is a widely held view of our

Ph.D. programs that they produce graduates who are more and more highly qualified for fewer and fewer jobs. And yet the great paradox is that we can all point to distinguished programs and successful graduates who are productive, satisfied, and gainfully employed outside of academe. As a community, one of our greatest challenges is to make the path to these careers more obvious and valued by the graduate programs.

My summary of themes that I heard at the Forum is certainly not exhaustive. But I think it is representative of areas that will continue to be discussed, and more important, areas that will be reflected in congressional deliberations.

I want to turn now to the budget, which also gives us important signals about what the future holds for NSF and the research enterprise. I would like to discuss what it foretells for future budgets. I consider the budget that the President announced last week as a prototype for future budgets. It contains four dominant elements that I believe we can expect to recur in future budgets:

1. Relatively modest growth.
2. Rigorous priority setting.
3. A set of activities I call productivity investments, which are closely connected to the National Performance Review.
4. Strong emphasis on research and education activities in strategic areas.

First, the growth rate. The President considers the NSF budget as an investment in the future and has provided NSF with a rate of growth of 6 percent, which in this budget environment is an outstanding increase. And I must add that there is no guarantee that Congress will support even this modest increase, especially given the deep cuts in programs that are funded by the same committees as NSF. (Engineering, by the way, is budgeted for a 9 percent increase, but again, it is very early in the cycle.)

There is also no denying that this increase is less than what NSF has received in recent years. For the past decade, the President's proposals for NSF have generally provided increases on the order of 15 percent. Even though Congress did not always fully fund these increases, NSF's budgets often grew by a rate of 10 percent or more.

I do not think we can expect to see increases of that order anytime in the near future. There are no peace dividends or windfalls to draw from other parts of the government that Congress can tap to provide a substantial boost for science. Instead of doubling scenarios, staying a few steps ahead of inflation is now an optimistic outlook. Therefore, I believe that 6 percent is likely to represent the upper end of budget growth we can expect to see through the turn of the century—provided we continue to demonstrate that science is a particularly good investment.

The second element of this budget that I feel is prototypical of future budgets stems directly from this modest growth rate: the will to set clear priorities. Priority setting is nothing new in a budget. The inherent purpose of budgeting, after all, is to allocate resources between different priorities. In the 1995 budget, however, the priorities across NSF are much more distinct than in past years because the overall resource constraints are so much tighter. We will have to stop doing some things in order to do others.

The starkest example of this priority setting is the funding of the academic research infrastructure program. This program provides funds for the renovation of academic research facilities and for the purchase of large-scale instrumentation. It was funded at $100 million for fiscal year 1994, but the proposed level for 1995 is just over half that amount, $55 million.

Clearly, these investments in infrastructure are badly needed. Virtually every campus can document a real need for laboratory renovations or replacement and new equipment.

But given the tight budget constraints for 1995, we made a difficult decision. We concluded that our incremental dollars would go farthest and do the most for the nation by funding activities in our research and education programs.

I do want to add that the Fundamental Science Committee of the President's NSTC will be discussing academic infrastructure very soon. And I am hopeful that we can develop a government-wide response to our academic infrastructure requirements in the not too distant future. In my opinion, the problem is of such an immense scope that the only way to address it is through a response involving all of the agencies that support academic research. Furthermore, the academic research infrastructure should not be viewed in isolation from that of industrial or federal laboratories.

I turn now to the third prototypical feature of this budget—what I am calling productivity investments. These include an increased emphasis on assessment and evaluation and investments in new technologies to streamline communication and the processing of proposals. Much of the framework for these activities comes from the National Performance Review—the effort Vice President Gore is leading to reinvent government.

A first example of what we are doing in this area is a pilot project, actually a number of projects, in electronic information dissemination and proposal processing. We know that this project will eventually save trees. But we also have great expectations that it will lead to changes that will streamline many parts of the process and save your time and your university's administrative costs. These savings will also accrue to NSF, Congress, and the agencies we are cooperating with in developing these projects.

We also have initiated an important set of assessment and evaluation activities. If you have a chance to read or at least skim through our full budget justification, you'll see much more discussion devoted to how we measure performance and track progress in key areas. On major construction projects we support, such as the LIGO [Laser Interferometry Gravitational Wave Observatory] project and the Gemini Telescopes, the budget now lists specific milestones for each year of construction through completion.

But I also want to make a point about what we are not planning to measure. I know that there has been a sizable amount of apprehension in the community about the recent emphasis on measurement and evaluation emanating from inside the Washington Beltway.

NSF will not be asking researchers to tell us on what day they plan to make a major discovery or when they will be 75 percent of the way toward that discovery or even if they will make an important discovery. No one has this in mind.

What we do have in mind is developing a common-sense approach to evaluation and performance measurement for the activities we support. I see it as being a process of experimentation—testing different ways of tracking progress, program accomplishments, and documenting results. Where these experiments work and add real value, we will incorporate them into our decision making and use them to set priorities. Where they don't work or create misguided incentives, we'll move on to other experiments.

But I hope all of us would agree that this process of experimentation is important and worthwhile. It will give us more evidence, and more convincing evidence, with which to document our contributions as researchers and educators. We are often criticized for relying too much on anecdotes to justify our worth. We need to approach the process of developing ways to gauge the value of our programs with the same rigor that we employ in our laboratories.

The fourth area that sets a prototype for future budgets is what I consider to be the dominant feature of this budget: support for research and education activities that address national priorities. As I said earlier, this issue was the focus of many discussions at the Forum.

In FY95, eight strategic areas receive a special focus in NSF's budget: high-performance computing and communications; global change research; advanced manufacturing technologies; science, mathematics, engineering, and technology education; biotechnology; advanced materials and processing; civil infrastructure systems; and environmental research.

These areas include the traditional FCCSET initiatives. For engineering, the focal points for research will be in civil infrastructure systems, advanced manufacturing technology, and Advanced Materials and Processing.

Our role begins, as I believe it must, with excellence and high standards. Through the use of investigator-initiated proposals and the merit review process, we adhere to the high standards for which NSF and the community it supports are known around the world.

Furthermore, in these priority areas, we take advantage of one of NSF's greatest strengths: its ability to foster connections of all kinds. Connections between research and education, between universities and industry, with other federal agencies, and across disciplines of science, mathematics, and engineering. These connections have proven beneficial to all participants, and to the progress of science, mathematics, and engineering as well.

Yet I am aware that NSF's focus on strategic areas is not warmly embraced by everyone. There is a sense that if NSF is supporting activities relevant to a national priority, then it is not being true to its core mission.

I understand such concerns, but I also believe they are overstated. I view this issue from a perspective of mutual benefits. We can identify many important, intellectually exciting areas of research that both advance the base of fundamental knowledge *and* are informed by needs in these priority areas.

I can think of many examples of research fitting this description. In my own field of atomic, molecular, and optical physics, there was never any question in my own mind or, I believe, in my colleagues' minds that much of our research was being supported because it fulfilled the mission of an agency, even though the work itself was entirely fundamental.

I want to close by noting the connection between the types of issues raised at the Forum and how these are reflected in the budget. It is difficult to know exactly how these two activities will play themselves out over the coming months. Nevertheless, it seems clear to me that they reflect a coalescence of opinion about post–Cold War priorities for science. We are developing a framework for setting priorities—a framework that continues to rely on the proven ability of the research community itself to submit for merit review its best ideas for discovery.

In the coming years the research community will be asked to focus its energies and its intellect more on areas of national interest. We will be asked to broaden our educational efforts and to seek better ways of increasing public awareness of the linkage between our work and national priorities. And we will be asked to provide clearer evaluations and a better accounting of the programs we undertake.

Challenges to the research community can be met in many ways. We can welcome the opportunity to make an even larger contribution in setting and responding to emerging national priorities, or we can seek to insulate ourselves from a world that is undergoing rapid and dramatic change. To me, the choice is clear. Our input is essential in identifying fields of inquiry where focused research will provide the basis for informed decisions, including those having to do with new technologies. I have every confidence that the research and education community will be invigorated by these challenges and will continue to seek a growing role in setting new priorities for the careful investments we must make to ensure the future strength of our nation and the well-being of its people.

A View from the Front Lines of Academic Engineering Research

Simon Ostrach

I will address the forces shaping academic engineering research today from the perspective of one who has been a working engineer for half a century. During this period the nature of engineering, business, and our country has undergone major changes. Even greater changes are indicated for both the near term and the next century.

Many commissions, committees, and boards have been established to address the major problems in industry, government, and academia due to the changing world, and many reports have been written and symposia and workshops held to indicate possible solutions. As I have read the reports and attended the forums, I have found that the views being expressed differed significantly from mine. For the most part, the participants and contributors to these activities were from the executive and administrative offices of industry, government, and universities, prestigious people, decision and policy makers. But where were the people who chose other career paths and continued to do technical work? Would their insights, like mine, be different from those expressed in the reports that were receiving most attention? Are perspectives from executive offices really so different from those from laboratories? It would appear that "working stiffs" are perhaps a neglected national resource, so I, with some trepidation, will try to represent them and present a different perspective on the subject. I have modified the subtitle of my paper because I did find at least two reports (National Research Council, 1987; National Science Board, 1992) that express views that are very similar to mine. I am puzzled as to why these have not received more attention.

INTRODUCTION

Until World War II each branch of engineering considered itself a distinct and separate field that was growing in an orderly evolution. It was expected that when students completed a college program in some branch of engineering, they were well prepared for their entire professional career. Their ancillary training in physics, chemistry, and mathematics provided little more than the basic principles of a given field. In effect, current practice was conveyed to such engineers, usually in the form of handbooks or correlations, which were used to solve problems. If the problem being considered was not identical to the known solution, the same formulas were applied but the safety (really ignorance) factors were increased.

The war imposed the need for sonar, radar, the atomic bomb, and many other applications that exceeded both the supply and the capability of engineers. Those products were developed, not empirically, but from basic principles, and physicists were largely responsible for them. This demonstrated rather dramatically that people with a good understanding of general principles could apply them to accomplish, rather quickly, ends that previously took long periods of time to accomplish by crude semiempirical methods. This wartime experience led a number of technological institutes and engineering colleges to believe that if fundamental and comprehensive knowledge of the physical sciences could so remarkably shorten the time from discovery to application, then the sciences and mathematics should receive greater emphasis in their education programs. Thus, in some schools new engineering education programs stressed fundamentals rather than current practice, and this culminated in what is now known as the engineering science curriculum. To further the scientific "coloration" of engineering (a phrase that can be attributed to Donald Frey, Northwestern University), engineering research was introduced in the universities and it has become an important part of the educational enterprise.

After World War II, as the United States assumed the role of a superpower, it was apparent that the nation's defense and economic and social well-being depended directly on engineering. The dominant feature of the environment in which engineering functioned in that period is *change*. The National Research Council's Committee on the Education and Utilization of the Engineer (1985c) identified four factors as particularly important in that regard for the engineering profession: (1) a large expansion in the roles of government, (2) a rapid increase in the amount of information in daily life and work, (3) the accelerating rate of technological development, and (4) the internationalization of business and the marketplace.

To date the engineering education system has been relatively successful in producing engineers able to cope with the changes, despite such severe constraints as departmental structures established in the nineteenth century, faculty shortages, uneven and declining student enrollments, obsolete equipment, and funding shortages. The adaptability of engineers to such changes has been attributed to the broad content of physical and engineering science in the current undergraduate curricula (National Research Council, 1985b).

However, despite much success, U.S. industry has experienced strong competition from abroad and has lost many markets in key products. Serious questions have been raised about how to counter those trends and, as a result, a plethora of boards, commissions, reports, symposia, and workshops have been organized around major examinations of the entire engineering profession.

The engineering education system has received its share of the blame, primarily because design and manufacturing (current practice) were not given enough attention in the curricula. Criticism has also been leveled at academic research: it is said to be only self-serving for the faculty to have publication records and that it is irrelevant for industry.

The end of the Cold War, with the associated reduction in defense budgets, the significant decreases or elimination of research in industry, and the policy that research must serve national goals—a policy promulgated by the new government administration with the support of influential members of Congress—all presage even more major changes for engineering research, in particular for such research performed in universities. Therefore, an examination of the nature of engineering research and its role in the profession and for the welfare of the nation is in order.

THE ROLE OF RESEARCH IN EDUCATION AND INDUSTRY

Research has become an important part of engineering education. Research is also performed in industry, more in some industries than in others. It is necessary to understand the purposes, goals, and types of research in these two different kinds of institutions before discussing possible synergisms.

Engineering Education

The principal responsibility of universities is to the students, their *primary customers*. The support that industry gives to universities is certainly helpful, but it is small in relation to the investment made by students (and their parents). Industry's support of research constitutes the smallest source of R&D expenditures at U.S. academic institutions (Dickens, this volume). In Dickens's view, the situation is somewhat better for "organized engineering research units at academic institutions," which receive almost 23 percent of their funds from U.S. business and industry. However, without data on the specifics of the disbursements, I suspect most of that support goes to just a few of the largest and most prestigious engineering schools. Thus, it is evident that industry is a user of university-developed products, in much the same way as the National Football League is. This clarification, however, in no way implies that meaningful interactions between industry and academe do not exist or are not desirable. On the contrary, it is merely intended to identify clearly the differences in objectives and functions of the two types of institutions.

The university, then, must transmit knowledge and understanding to young people, give them an opportunity to develop their capabilities, help them gain an understanding and appreciation of the world around them, teach them to think independently, and in some curricula, like engineering, enable them to obtain skills that will serve them well throughout their lives.

We have seen that before World War II the skills transmitted to engineering students were those of "current practice," which were based on lore, empiricisms, and intuition. It should be kept in mind that it was during that period that many industries were developed by those means. The wartime lesson that the time from discovery to application could be considerably shortened by fundamental knowledge and research led to major curricular changes that deemphasized current practice, including design and manufacturing. The new curricula did produce engineers who were flexible, versatile, and adaptable and who functioned well during the many and rapid changes that have occurred in engineering since the war.

Engineering research was intended to confront a student, for the first time, with a complex problem that was not well specified, would need defining, and would require synthesis of all the student's knowledge for its solution. In this way the student would experience the loneliness of individual inquiry and the anxiety of the unknown and would develop the discipline, tenacity, and perseverance required for exploring the unknown and for independent thinking. This type of research requires persistent work for a period of several years before the crucial insights and results are obtained. Significant advances have been made in this way, but the main purpose is to provide new and unique educational experiences for the students.

The situation portrayed above would seem to present a clear picture of what is needed for the future: more engineering science education and research. However, there are serious shortcomings to that approach if one desires the products of such an education to be gainfully employed, contribute to meaningful technological developments, enhance an employer's productivity, or help the nation's economic growth.

The engineering science curriculum in attempting to emulate the pure sciences and thereby gain academic respectability, developed courses that were analytic, formalizable, and teachable. Thus, well-posed problems were presented and emphasis was given to solution methods and their results, which sometimes illustrated interesting physical phenomena. The problems considered were chosen primarily for their mathematical tractability and, thereby, represent highly idealized situations with limited, if any, relation to real engineering systems. Furthermore, the conditions under which the simplified results could be applied to real problems were rarely, if ever, delineated. Although such criticism is mostly made about theoretical studies, it applies equally to experimental work that is designed for ease of observation and measurement rather than its relation to real situations. Much of the academic research has this character as well. It is thus, perhaps, not surprising that academic research is said to be "pure" or "basic" and is relegated to the "ivory tower" and considered useless by industry.

Also, this approach has deprived many students of one of the most essential of all engineering skills, the ability to determine a priori, the essence of a complex situation, that is, to define the meaningful problem. More emphasis is required for problem definition, and consideration of open-ended real-world problems, the type of interest to industry, is urgently needed.

Industry

To determine the appropriate role of engineering research in industry, it is first necessary to recall some changes in engineering practice that have occurred in the past half century. Almost all industrial and manufacturing processes were developed empirically, as was the related equipment. Throughout the years, those processes remained essentially unchanged as long as the companies were profitable and the industries were unchallenged. This is true not only of traditional, heavy industries, because many high-tech industries also have empirical origins and processes.

As major U.S. industries began to experience strong competition from abroad, it was suggested that matters could be improved by the use of computers and robots. In fact, in numerous studies of changes in engineering practice, new engineering tools based on the computer are said to be part of a revolutionary change in how engineers work (see, for example, National Research Council, 1985a). Thus, the popular buzzwords associated with modern engineering are terms such as robotics, CAD/CAM and CAI/CAP. The improvements made in this way are most welcome, worthwhile, and overdue. However, it must be understood that, for the most part, the same basic elements of the system (machinery) are employed, albeit faster, more accurately, and more uniformly.

The deemphasis or elimination of design and manufacturing in engineering education is sometimes said to be a contributing factor in the loss of industrial competitiveness. Much pressure is being applied to increase emphasis on those subjects in engineering schools, and considerable federal support is being given to programs dealing with those subjects. What does not seem to be recognized, however, is that all the changes in engineering have also changed the nature of design and manufacturing, that is, engineering practice has changed. It does not seem to be fully recognized that with demands for products with greater purity, ability to withstand more severe operating conditions than ever before, and greater precision and economy of manufacturing, there is little or no experience or knowledge on which to base such designs. In examining various industries, it is readily apparent that there is a large margin between actual existing industrial systems and the limiting physical behavior, as determined by the laws of nature. Thus, there is great potential for improving industrial and manufacturing productivity by enhancing the effectiveness of the related processes. To accomplish this, it is necessary to "research" the industrial processes, that is, to gain an understanding of the phenomena involved in the process and the factors on which they depend. Vigorous and comprehensive engineering research programs that are directly related to real problems are necessary to develop the knowledge base and physical principles on which advances in design and production can be based. In so doing, gaps in existing knowledge are identified for further study and, also, it is readily apparent that such research is essentially cross-disciplinary.

A National Academy of Engineering (NAE) study committee expressed a similar view: "The technical intensity of most manufacturing and service industries will continue to grow at an accelerating pace, and commercial technology will become increasingly science-based and interdisciplinary" (NAE, 1993, p. 92). The implication of this statement is that R&D activity must be pushed "further downstream into design, production, and marketing, as well as factoring production and marketing considerations into the earlier phases of upstream development activities" (NAE, 1993, p. 31).

It might appear impossible to deal with the complex and diverse phenomena that occur in industrial processes. In fact, most industries either feel no need to apply new knowledge or think their processes are too complex for detailed study and so depend on empiricisms and gross correlations. On the other hand, much academic research is too specialized or idealized to be of much value to industry. Thus, there is now a need for new and intimate relationships between industries and engineering schools so that there can be a coupling of technology with all the latest developments of engineering research, such as the increasing power of theory and computation, meaningful model systems, and sophisticated measurement and diagnostic tools. Such university/industry relationships should not be expected to yield, for example, a generalized computer code that will solve all the company's problems, an approach that is, unfortunately, being pursued too frequently.

The research being advocated here is fundamental engineering research. This is distinguished from fundamental science research in many ways. For example, science research primarily seeks new knowledge about the natural world without regard for its utility. Engineering research focuses on the man-made world in order to expand the knowledge base and to identify and exploit the physical principles on which advances in design and production can be based (National Research Council, 1987). In many cases there are interactions between science and engineering research, and the boundaries between them are often difficult to discern. However, basic engineering research that provides the underlying competence on which applications or applied research is based is often cross-disciplinary, whereas basic science research is mostly constrained by scientific disciplines. Some basic engineering research does not directly involve the laws of nature but addresses the functional characteristics of large systems consisting of intricate components. The knowledge base for manufacturing, for example, will ultimately consist of engineering principles drawn from many engineering disciplines and activities. There seems to be general misunderstanding of these distinctions, as is evidenced by the usual relegation of basic research to science and applied research to engineering.

The National Science Board (1992, p. 47) stated this crucial issue in the following way:

> A pervasive problem in the United States today is the insufficient attention given to fundamental engineering research in industry, government, and the universities. Every firm must have an ever-expanding, relevant engineering knowledge base, and the hardware and software techniques for translating that base quickly into practice, in order to convert ideas into products rapidly and efficiently. The often-cited lack of emphasis on process improvement and manufacturing, along with excessive time delays from concept to available product, attest to a pervasive lack of understanding of, appreciation for, and sufficient attention to the vital role of fundamental engineering research by U.S. industry, government, and universities. Yet, there is no sufficiently broad and deep fundamental engineering research base on which to build; furthermore, there are an inadequate number of engineering researchers in U.S. industry who are equipped to, and called upon to, extend that base as needed. The greater the storehouse of fundamental engineering research, and the greater the ability in industry and government to extend it as needed for proprietary or national reasons, the better able the qualified engineer is to innovate in an integrated system of design, manufacture, and maintenance.

That report also says that too little support is given to process-oriented R&D. U.S. industrial R&D is weighted much more heavily toward product technology than process technology. In relation to their Japanese counterparts, U.S. firms also allocate a disproportionately small share of their R&D budgets to the search for new or improved processes.

UNIVERSITY/INDUSTRY INTERACTIONS

The perspectives presented above provide a basis for meaningful interactions between industry and academe. However, some observations need to be made first to gain an appreciation of why such relationships have not developed to the degree required.

Current Views

Birnbaum (1994) states the university's position as follows: "Unfortunately, it has become all too common to place the onus for the supposed failure of basic research to contribute to economic competitiveness on the basic research sector."

At a recent meeting on "World Leadership in Basic Science, Mathematics, and Engineering" sponsored by the Office of Science and Technology Policy, there was throughout a tacit assumption that industry was waiting impatiently for research results, which were not forthcoming. That is contrary to my experience and does not seem to be representative of industry's position.

> Basic research has not failed—it has fulfilled its mandate. The universities have not failed; if anything, they have succeeded too well in educating large numbers of excellent scientists and engineers. What has failed is industry's ability to translate the fruits of basic research into products and profits. The reasons for the failure are manifold—too numerous to enumerate at this point. It is important to point out that the source of failure of industry to capitalize on available basic science is primarily a failure of management and not a failure of the scientists and engineers (Birnbaum, 1994).

Concurrence with this conclusion from the industrial side seems to be presented by Armstrong (1993, p. 5), who states that "responsibility for deficiencies in our industrial performance rests largely with failures in the private sector, failures of strategy, investment, and training—in short, failures of management." However, the polarity between the university and industrial positions, or perhaps between the view from the executive office and the workbench, is well illustrated by Armstrong's further remarks:

> These (failures) will not be cured, or even helped by more research. Trying to cure poor industrial performance in the short term by more university research is like asking for helpers when pushing on a rope. . . . Poor technology transfer from the university or national tabs to industry has not been a major cause of our competitiveness problem (Armstrong, 1993, pp. 5–6).

It is, thus, not surprising that industry has largely abandoned basic research. Armstrong gives emphasis to the short term and indicates many other factors involved in industrial competitiveness, such as fiscal strategies and policies, marketing, the economic climate, and trade policies. He then states that "it is fair and accurate to say that universities lack deep understanding of products or markets, have no responsibility for development or manufacturing, and tend to overestimate the importance of science in technology competitiveness" (Armstrong, 1993, pp. 6–7). All this is valid, since competitiveness is a highly complex combination of attributes. However, as indicated above, one of the primary skills of well-educated engineers is that they can extract the essence of a complex problem. From that vantage point, I find that the greatest leverage that can be applied, by industry itself, is to "research" its processes (the heart of the endeavor) to improve, modify, or replace them as necessary to bridge the chasm between existing, empirically derived processes and the possible improvements that are more efficient and effective. Orders-of-magnitude improvements in time required to complete a process or in the quality of the product, or both, are possible. Surely, such developments could tip the scales in global competition. Unfortunately, too many executives and policymakers seem to be unaware of the power of technical solutions to industrial problems.

Other arguments diminish the role of research in industry. The Committee on Science, Engineering, and Public Policy (1993) reported that for "industries that rely on high technology but are technically self-contained (such as the semiconductor industry) and industries that do not depend heavily on current science (such as the automobile industry), the results of current fundamental research are generally not decisive." I am not sure what is meant by "technically self-contained," but I do believe both those industries have large margins for technical improvements. That report goes on to point out that Japan, which is not a leading research power, does very well in such industries by "strategies largely separate from scientific research, but highly dependent on engineering." Japan's industries developed after the war in an era when engineering was evolving away from empiricisms and did not carry the baggage of capital investments from another era, as does the U.S. industry. Since research results are not constrained by national boundaries, the Japanese used them freely to send back to us improved products. Whether what they used is basic research or "engineering" is arguable. The fact is that they started fresh with mid-twentieth-century knowledge and an openness of mind to try new ideas, which is in sharp contrast to U.S. industry.

Future Needs

Obviously, there are different views and opinions about the role of academic research in support of industry's needs and the nation's welfare. To define meaningful university/industry relations, the boundaries of participation must be delineated. The major purpose of academic research is, and should remain, as an essential element in the education of engineers who will later do or supervise the high level of engineering required by industry and the government, including the "proprietary" fundamental and applied research needed when the knowledge base is inadequate. Academic research done in cooperation with industry will be of mutual benefit when both know their respective roles and are prepared to learn from each other. Exhorting universities to do more "relevant" research is likely to be counterproductive. It is unlikely to move an industry forward when that industry does no such relevant research and, more important, when it does not employ a sufficient number of highly educated engineers to use existing relevant knowledge and extend it as needed.

As pointed out by the President's Council of Advisors on Science and Technology,

> Some of the cultural differences that have long surrounded industrial research and university research have had the unfortunate effect of unnecessarily inhibiting the most effective interaction between industry and universities. The notion that each sector had its own well-delineated and isolated role and that new knowledge would flow as rapidly as necessary in one direction from the university to industry is completely at odds with today's world . . . . Despite recent gains in building links between U.S. universities and industry, there are still too many individuals in each sector who hold negative perspectives, attitudes, and stereotypes with respect to the other sector. The nation cannot afford to have this situation persist, and much more effort is required to overcome it. Even fundamental research that is not expected to yield short-term answers to industry's problems can benefit from being informed by the technical concerns of industry. Conversely, U.S. industry should have the benefit of easy and immediate access to new knowledge and new talent generated by the universities.

A couple of the relevant "cultural differences" require comment. The time scale for academic research is on the order of years, whereas industry looks for answers in periods of months. Such a mismatch must be acknowledged and addressed in any good partnership. Many of industry's activities are multidisciplinary in that they involve many people other than engineers, such as economists, lawyers, managers, marketers, and the like. Therefore, teamwork is an important and desirable mode of operation. As a result, there are increasing pressures on engineering schools to give "team" experience to the students. This is certainly worthwhile and needs to be done. However, it is being suggested that research also be done by teams, because it too must now be more interdisciplinary. However, *inter-* or *cross-disciplinary* means across academic or professional disciplines, and it is different from multidisciplinary.

Independent research develops abilities and qualities that are vital to industry, government, and universities. Engineers with such experience will play an increasingly important role in technical problem solving. Good (1993) states that "In the future, the complexity of engineering design tasks will require engineers with a doctorate degree." Many important discoveries have been made by a small number of very gifted people who were given the opportunity and time to pursue their ideas and intellectual interests. Therefore, independent research must be continued, supplemented by an educational program that emphasizes cross-disciplinary subject matter.

The fact that industries have significantly downsized their basic research or abandoned it completely indicates that a natural role for universities is to carry out the basic research required for industry and that industry focus on applied research and product development. In fact, engineering education would be enriched by consideration of real problems. However, the reasons for the abandonment of basic research by industry must be understood. Obviously, research is not deemed essential, so is industry even interested in interacting in a meaningful way with universities? Also, it is clear that fiscal support cannot come from the universities, and it probably will not come from industry, although it would be less costly than in-house work. Therefore, the federal government will have to be involved, which brings in its own set of problems. Engineering research is an essential area of technical activity that is seriously undersupported in the United States. As the National Research Council's Engineering Research Board wrote in 1987,

> This research is essential because all creative technological development in an intensely competitive world rests on it; yet it is undersupported because its central role in the development of productive goods and services is not clearly understood or recognized.

Despite the recent awareness of the increasing cross-disciplinary nature of engineering research, there is little overt support for such activities.

SUMMARY

From the perspective of a working engineer, I have pointed out a number of aspects of the changing nature of engineering that do not seem to be widely recognized and that directly impact the matter of academic research in a changing world. In particular, "engineering practice" has changed so that time-consuming empirical approaches are no longer competitive. Because technological advances have surpassed general knowledge, research is now required to develop a knowledge base for design. What is required is essentially cross-disciplinary, basic research, which is different from basic scientific research. Technical solutions to the problem of industrial competitiveness require more process-oriented research.

Academic engineering research done independently by individuals is an essential element of the educational process. Emphasis on problem solution rather than problem formulation is a deficiency of modern engineering education. Consideration of open-ended problems of importance to industry would enrich the education process.

Differences in viewpoints exist between academics and industrial people and between executives, administrators, and managers, and working-level people, just as "cultural" differences exist between academe and industry. If the necessary partnerships are to develop between the two sectors, then those disparate views must be addressed to find the bases for accord. Such dialogues should involve people from all groups, industry, universities, and government, and from all positions, executives, administrators, managers, and particularly, working-level people, who seem to be very much underrepresented.

REFERENCES

Armstrong, J. A. 1993. Research and competitiveness: Problems of a new rationale. The Bridge 23(1):3–10.

Birnbaum, H. K. 1994. Research Interactions: University, Industry, National Laboratory. Briefing paper for the Forum on Science in the National Interest, World Leadership in Basic Science, Mathematics, and Engineering, Executive Office of the President, Office of Science and Technology Policy, January 31 - February 1, 1994, Washington, D.C.

Committee on Science, Engineering, and Public Policy. 1993. Science, Technology, and the Federal Government: National Goals for a New Era. Washington, D.C.: National Academy Press.

Good, M. L. 1993. Industry needs and the curriculum. Issues in Engineering Education, Vol. 2, October. Washington, D.C.: Board on Engineering Education, National Research Council.

National Academy of Engineering. 1993. Mastering a New Role: Shaping Technology Policy for National Economic Performance. Report of the Committee on Technology Policy Options in a Global Economy. Washington, D.C.: National Academy Press.

National Research Council. 1985a. Engineering Education and Practice in the United States: Foundations of Our Techno-Economic Future. Report in the series Engineering Education and Practice in the United States, by the Committee on the Education and Utilization of the Engineer, Commission on Engineering and Technical Systems. Washington, D.C.: National Academy Press.

National Research Council. 1985b. Engineering Employment Characteristics. Report in the series Engineering Education and Practice in the United States, by the Committee on the Education and Utilization of the Engineer, Commission on Engineering and Technical Systems. Washington, D.C.: National Academy Press.

National Research Council. 1985c. Engineering in Society. Report in the series Engineering Education and Practice in the United States, by the Committee on the Education and Utilization of the Engineer, Commission on Engineering and Technical Systems. Washington, D.C.: National Academy Press.

National Research Council. 1987. Directions in Engineering Research: An Assessment of Opportunities and Needs. Report of the Engineering Research Board, Commission on Engineering and Technical Systems. Washington, D.C.: National Academy Press.

National Science Board. 1992. The Competitive Strength of U.S. Industrial Science and Technology: Strategic Issues. Report of the Committee on Industrial Support for R&D. Washington, D.C.: U.S. Government Printing Office.

President's Council of Advisors on Science and Technology. 1992. Renewing the Promise: Research-Intensive Universities and the Nation. Washington, D.C.: U.S. Government Printing Office.

Reengineering the Academic Engineering Enterprise

Chang-Lin Tien

This is a time of unprecedented change as powerful forces are reshaping the world around us. In this era of change, the new challenges and opportunities for engineering are breathtaking. Four major forces are transforming our society, and they are posing major challenges for our engineering research. The first is the political force of democracy and international peace—a powerful force, indeed. For nearly a half century, concern over national and international security dominated the federal drive to build the academic research enterprise. Then, almost overnight, the Cold War has become a chapter in history books.

Now that national defense and space are no longer the foremost priorities, the national research agenda is a matter of debate. This debate is intensified by the call for greater accountability in research and the funding tug-of-war between "big" science and "small" science.

The second force is internationalization of the world community. In the global village, national borders are fading quickly. The intense competition in the global marketplace is hitting American businesses hard. As Japan and other nations have imposed numerous obstacles to U.S. entry into their domestic markets, there are legitimate concerns that the playing field is not level. Nonetheless, in response to international competition, many major U.S. corporations have made a short-term correction that holds serious implications for the future. These corporations are scaling down their great industrial research laboratories or phasing them out altogether.

The force of internationalization affects us in other ways as well. As citizens of the global village, we are bound together by our common interest in protecting the environment. No single nation can solve problems that are global in scope, and no single nation can escape them. If we are going to survive, we must work together to save the air we breathe, the water we drink, and the food we eat.

The third force sweeping the world is the massive movement of people. There are 18 million refugees in the global village today. Another 24 million people are displaced inside their own countries. This means a total of 42 million people have been forced from their homes because of hunger, natural disaster, war, and persecution.

Today there are very few nations that do not send or receive international migrants. Massive migrations have long-lasting effects on nations and their economies, social institutions, health, environment, and relations with other nations.

I am part of this international movement. My family fled China when the Communists took over. First, we settled in Taiwan. Then we were fortunate to make our home in the United States.

Part of America's demographic transformation is a by-product of international migration—a transformation that deeply affects the academic engineering enterprise. A new wave of students will flood American colleges by the middle of this decade. In contrast to the young people who poured into our institutions two decades ago, these students will be a highly diverse group—diverse in culture, race, religion, income, and language.

The information revolution is the fourth force reshaping our world. Advances in communication technologies are changing the way we do business and even the way we conduct our daily lives. The couch potatoes of tomorrow will be "television users," not "viewers." We will transact purchases through our teleports. We will hold "video" meetings with friends and colleagues around the world. And, when we want to relax, we will match our wits in video games against opponents down the block or across the continent.

As these forces transform our society and our world, the academic engineering enterprise has not moved with the speed necessary to respond. As a result, the public has started to question the value of engineering technology. In the first part of this century, advances in engineering technology were regarded as essential to the prosperity and progress of our nation. Today, all too often people view engineering as part of the problem, not the solution.

Spectacular disasters have reinforced this view. People are not likely to forget "Three Mile Island" or "Challenger" or "Hubble" in the near future.

This era of change poses major challenges for the American academic engineering enterprise. This is not the time to cling to the status quo. Great demands require bold action.

We must reengineer the academic engineering research enterprise. We must direct our resources to meet new challenges. We must find ways to take advantage of the rapid transformation in the world around us.

In the last century, this nation has forged the most productive academic research enterprise in history. It is clear the challenges ahead will test this marvelous enterprise. But I am confident we will meet the challenges, and we will enter the twenty-first century as a world leader.

I want to propose some ideas on how we can reengineer the academic engineering enterprise. First, I would like to discuss how we can maintain and improve the pipeline of engineers. Second, I would like to suggest some ways to stimulate the basic and applied research that is so vital to the future of our nation.

Let me elaborate. First, we must prepare for the new wave of students that will be entering colleges and universities in the mid-1990s. We must act quickly to strengthen the pipeline of engineers.

Some may question how a discussion about the pipeline relates to academic engineering research. Yet I believe this issue is central. Unless we build the pipeline of engineering talent, we will not have first-rate engineers to lead a world-class research enterprise in the new century.

The leaks in the educational pipeline are well documented—not just for women and minorities, but for all students. Perhaps the most revealing study of all found that the longer American students are in school, the less they like science and math. Clearly, we are not doing enough. It is time for us to try courageous measures that will reach students starting in elementary school and continuing through the postdoctoral level.

Let me suggest why our attempts to solve the pipeline problem have fallen short. Instead of shoring up the pipeline, we continue to block it at several key points.

What do I mean? Our profession has a very rigid notion about how you become an engineer. We put lots of obstacles in the path of potential candidates. We don't pay enough attention to offering the kind of teaching and curriculum that excites and involves students. Too often we require competition for the sake of competition.

Only those who are both dedicated and adept at clearing these obstacles will succeed. Yet what about those who fall by the wayside? Are they any less talented than those who make it to the end of the line? All too often the answer is no. We lose exceptional students who find other fields of study to be more fulfilling.

It is not enough for us to agree that we have leaks in the pipeline. We must be brave enough to discard outdated notions about what makes an engineer. We must try to recruit and retain all kinds of people to our field—whether they are women or men . . . whether the color of their skin is brown, black, yellow, or white . . . whether they thrive on competition or not. We must remove the obstacles in the pipeline, not add to them.

Let me make it clear that I believe we must continue to set very high standards in engineering. That should not change. But in addition, we must help people to make their way through the pipeline, not get in their way. As we help them, we can make sure they meet rigorous standards.

There are four critical points in the pipeline where we are losing students. Students start to lose interest in science and math fields in secondary schools. If students fail to fulfill math and science prerequisites at the secondary level, they find it extremely difficult to catch up. Few are likely to enroll in college engineering programs, and fewer still are likely to succeed. There are some successful programs in elementary and secondary schools aimed at improving math and science learning. These programs need more resources to do a better job of reaching more youngsters.

Engineering education at the undergraduate level demands our special attention as well. We lose many of our brightest undergraduates to the social sciences, humanities, and other professional schools. Studies show that as many as 60 percent of undergraduates in engineering, science, and math switch to other fields.

This is especially a problem for women and minority students. Women receive about 15 percent of the engineering degrees in American universities. Although this marks an improvement, it is still a relatively low representation, considering that women earn more than half of all bachelor's degrees. The representation of African Americans and Hispanics is lower still. Altogether they represent less than 10 percent of all engineering graduates.

The next leak occurs at the graduate level. A growing number of engineering degree-holders are going into business, law, medicine, and other fields. Graduate engineering programs rely more and more on international students. Although international students increase the pool of talent and help fulfill the people-power needs of our nation, the vast influx has serious implications for our engineering practices and culture. We must study this trend more carefully to better assess the effects on our engineering enterprise.

The leaks in the engineering pipeline continue all the way to the faculty level. Nowadays some of the most promising young engineers are choosing industry and business over academia.

Let us discuss how we can stop the leaks in undergraduate programs and at the faculty level. These are major concerns that must be addressed by the academic engineering enterprise.

What can we do to attract and retain the most talented students in academic engineering programs? First, schools of engineering must listen to our students and design the programs that meet their interests. Not surprisingly, student interests reflect the changes in our world. Students today are less interested in defense-related fields, while more are entering fields associated with the information revolution, environment, and biotechnology.

The trend in student applications for freshman admissions at the University of California, Berkeley, is a case in point. Electrical Engineering and Computer Sciences continues to be the most popular department in engineering. Can you guess which field ranked second in popularity? Not civil engineering. Not mechanical engineering. Even though these are top-ranking departments at Berkeley, the number of applications to bioengineering surpassed them. Environmental engineering and general engineering science were popular as well.

Engineering faculty and curricula must be flexible so we can offer the kind of undergraduate majors that reflect both the interests of students and the changing forces in the outside world. The first step is to offer the kind of programs that reflect both student interest and real-world demands. Yet this is not enough. The second step is to provide the kind of instruction and support that will maintain the interest of students and pave the way to academic success.

Let me cite a couple of examples from Berkeley. It is a common complaint about engineering programs that in the first two years of college, most students do not have time for courses that give them a taste of engineering. Instead, they are struggling to get through all the prerequisites. It is no surprise that many lose interest before they enroll in their first engineering class.

We must give students a taste of engineering right from the start. Indeed, for a few years, I volunteered to teach an introductory course on engineering at Berkeley High School.

Berkeley's College of Engineering is taking this approach. An exciting new seminar introduces freshmen to the field of engineering. Professors who are authorities in different fields lecture to the class. For instance, Professor Abolhassan Astaneh, who is working with the State Department of Transportation to seismically retrofit bridges throughout California, took students by boat to the Bay Bridge. He pointed out the damage from the Loma Prieta earthquake in 1989 and the retrofitting measures that he designed. Later, students used computers to design a bridge and estimate its cost. Classes like this generate considerable excitement in the field and keep undergraduates in the pipeline.

Another effective way to build the pipeline is to foster teamwork. It is part of our tradition in engineering to take pride in encouraging the kind of competition that separates the wheat from the chaff. But if we look a little closer, we will find that we are losing bushel loads of wheat as well.

Some engineering professors assign students to work in teams and guide them as they learn to collaborate on projects. After years of competing for grades, most students are accustomed to working on their own. They must learn how to work together effectively. Not only are team projects a good way to learn, they also are sound preparation for most careers in engineering.

Yes, it is essential that the academic engineering enterprise focus on the undergraduate educational experience. It is just as important for us to stop the leaks at the faculty level.

First-rate academic engineering programs depend on recruiting and retaining the most promising degree-holders. One problem is the disparity of salaries between the business world and academia. For instance, it is not unusual for recipients of engineering doctorates to look forward to earning as much as $100,000 a year at investment houses.

There are other reasons as well for the growing number of defections at the faculty level. Often the frustrations associated with starting up a lab and maneuvering a complex tenure system are deterrents to young engineers. Winning grant support is extremely difficult. Today the competition is so intense that major federal sources fund only 25 to 30 percent of all grant proposals. For starting junior professors, the success rate may be significantly lower.

Without grant support, the new professor cannot maintain an active lab. Without an active lab, the new professor cannot conduct the kind of research that will lead to tenure. So, instead of climbing the academic ladder, many talented new professors find themselves trapped in a vicious cycle. It is hardly surprising that too many top young engineers are turning to industry and the business world.

There are no easy answers for reversing this trend, but let us consider the following proposals. First, we must urge the National Science Foundation, National Institutes of Health, Department of Defense, and other major funding sources to build grant and fellowship programs that are aimed specifically at researchers who are in the early phase of their careers.

Second, all of us must do more to help—not hinder—young professors in their climb up the tenure ladder. Again, just as in the case with undergraduates, competition for the sake of competition does not necessarily make sense.

Senior faculty need to mentor younger faculty. We should advise them on possible avenues of research. We should help them secure grant support. We should encourage them to publish. We should help them to balance demands in teaching and research.

We should consider team research efforts as well. Interdisciplinary projects that involve both senior and junior faculty offer many advantages. However, in team projects, we must make sure that junior faculty develop on their own.

Clearly, our mission in building the engineering pipeline is demanding. Now let me turn to our mission in research.

I believe we need to do a better job of channeling the forces that are changing our world in order to fulfill our mission in research. I want to discuss some ideas for succeeding in the global arena and taking advantage of new communication technologies. Then I would like to turn to some proposals for the national agenda in basic and applied research.

First, our engineering enterprise must do more to promote interaction in the global community. Proficiency in the English language and familiarity with the American culture are no longer enough for success in the international arena. Advanced studies and overseas appointments are one of the finest ways to gain insights about other cultures as well as learn about advances in the field. Yet many young engineers are reluctant to pursue overseas opportunities.

There is a serious concern that international businesses take advantage of American academic resources. Yet we should be just as concerned that we are failing to take advantage of resources offered by other nations.

Consider the example of Japan. Despite the availability of grants and appointments—enough in Japan alone to fill two National Science Foundation directories—many young American engineers are discouraged by differences in language and culture. Moreover, many regard these opportunities as interruptions in their careers, a view that all too often is reinforced by their institutions.

This reluctance to study and work in Japanese institutions means the U.S. engineering community is losing valuable insights into fields where Japan is gaining dominance. Although Japanese engineers who come to the United States face language and cultural barriers, they are highly motivated. There is an unwritten rule in major Japanese universities that engineers and scientists cannot advance to full-professor status unless they have conducted research or postdoctoral work in the United States or Europe.

American institutions should offer incentives to our young engineers as well. We should encourage them to take advantage of overseas opportunities so that all of us can benefit from the new knowledge developed by other nations. We should reward our young engineers for pursuing opportunities in Asia and Europe that will help them develop lifelong professional contacts. It is not by chance that all of my recent Ph.D. students do research or postdoctoral work in Japan and Germany. They know I will give them top recommendations when they take advantage of research fellowships and visiting faculty positions in these countries.

As a matter of fact, one of my Ph.D. students who has just earned his doctorate is appointed to be an assistant professor at Tokyo University. Although it is still unusual for an American Ph.D. to receive a faculty post in a Japanese university, this appointment shows the growing internationalization in academic engineering programs.

Earlier I mentioned how the information revolution will affect our personal lives. I believe we must take full advantage of interactive TV and other new communication technologies in our research as well. We must become aggressive commuters on the information superhighway.

The potential is fantastic. Interactive television opens opportunities for new modes of collaboration in research across the country and around the world. For instance, we could hold regular video conferences with colleagues instead of having to travel around the country and world every time we need to collaborate.

Already telephone companies and major corporations are investing hundreds of millions of dollars in market tests of interactive television services. The academic engineering research enterprise must stay in the loop. We must conduct in-depth discussions with corporations about our instructional needs and explore together the potential uses of interactive services in our laboratories and classrooms.

Telecommunications advances also pave the way for more rapid and effective dissemination of information. The academic engineering research community already relies on electronic library indexes and electronic billboards. Now the most established journals are investigating how to get on line.

It is critical for us to make full use of electronic journals and information databases that can be readily accessed by users worldwide. "Publish or perish" has been the credo of academic researchers. By the new century, perhaps we should say instead, "Get on-line or face decline."

Taking advantage of the information revolution and becoming successful players in the global village will help us in our research. Now let me turn to research itself.

How can we stimulate highly creative basic and applied research? First, we must do a better job of encouraging high-risk research by individual professors. Although our peer review system has been highly successful in the last 30 to 40 years, we are starting to see a disturbing trend. A growing number of engineers and scientists are sticking to safe research that leads to incremental advances. This is not by chance. Research projects based on conventional thinking are far more likely to win the support of reviewers than high-risk research proposals.

To address this problem, I propose that federal funding agencies consider a pilot program. This pilot would provide 1 percent of the agency's funding total in the previous year to create a pool for creative, high-risk projects. Each university that receives funds from this pool would be required to respond with one-to-one matching funds. Both universities and funding agencies would monitor and evaluate these efforts closely.

We must encourage applied research as well. In the 1950s and 1960s, engineering science enjoyed a tidal wave of popularity. Many American engineers and scholars started to concentrate their research on physical phenomena and mechanisms. New and improved devices, designs, and manufacturing processes have received far less attention. This failure to apply valuable knowledge poses a serious problem for our nation.

I believe the academic engineering enterprise and industry must forge stronger partnerships. The working links start with the individual professor. As engineers, we need to go into the real world and solve real problems.

I want to draw from an observation of Professor David Patterson, who heads Berkeley's Computer Science Division. In a recent industry publication, Professor Patterson said he carefully selects a problem to research. Then he collaborates with industry while doing his research. His last step is, and I quote, "doing the missionary work to convince people to use the ideas."

This is a method that works. Professor Patterson collaborated with Stanford University and Silicon Valley researchers to develop and demonstrate the RISC technology. Today this computer chip design is widely used in the computer workstation industry because it has increased performance and lowered costs.

Yes, it is important for individual professors to have working ties with industry. Engineering schools need to develop strong links with business and industry as well. The traditional approach is the industrial liaison program, which has been successful on campuses across the country. We need to build on this effort, trying more innovative approaches to forging links with business and industry.

As some U.S. businesses and industries scale down their research operations, it is also important for us to explore partnerships that go beyond the confines of industrial liaison programs. The chief executive officers of industry, the presidents of American universities, and the directors of federal laboratories must meet together not once, not twice, but as part of regular roundtable discussions.

The success of basic and applied research in the United States is unduplicated anywhere in the world. But our nation falls down when it comes to putting this knowledge to work.

We need regular exchange if universities are going to pursue avenues of research that can be applied. We also need open interaction if industry is going to take advantage of the new knowledge generated in university and federal laboratories.

Industry and academia can benefit if the private sector plays a larger role in supporting and participating in academic research. It is ironic that many leading U.S. academic research laboratories have received more offers of funding support and visiting scholars from Japan and Europe than from American industries. It would benefit all of us if U.S. industries moved ahead of our international competitors and interacted more with academia.

This means we must develop clear guidelines to avoid conflicts of interest, threats to academic freedom, and undesirable forms of foreign participation in our research enterprise. Only if the leaders of business, industry, and academia take part in roundtables can we develop the kind of exchange that will lead to more productive collaboration in the future.

The federal government should play a stronger role in promoting applied research as well. With the decline in private-sector research, the federal government should consider taxes and other incentives aimed at encouraging new forms of engineering research collaboration among universities, corporations, and national laboratories.

University researchers should become better missionaries—a notion I am borrowing from Professor Patterson. We must be persistent and forceful in persuading our peers in academia and industry about the potential economic and social value of new knowledge.

Meanwhile, industry must be willing to make the investment of time and resources. New technologies cannot be brought to market overnight or even in a year. Industry must be willing to make the commitment of a decade or more. Corporate goals and resources may need to be redirected as well. This was the case for GE—a world leader in medical imaging systems. This was the case for Motorola—a leader in pagers and cellular telephones. And, this was the case for Corning, a leader in fiber optics.

Both industry and academia have our work cut out for us. Only if both sectors fulfill our responsibilities will the United States gain a larger share in the international market and continue to be an international leader in the twenty-first century.

These are challenging times for all of us. Yet these challenges open up many opportunities. So let us take heart and mine the many wonderful opportunities in our world today.

Defense Budgets and Academic Research

Duane A. Adams

With the end of the Cold War and the resultant pressures to downsize defense programs, there are real concerns about the impact that Department of Defense (DOD) spending will have on academic research programs. The procurement accounts have been cut in half over the last five years, but the defense leadership has been committed to maintaining a strong science and technology program. The Advanced Research Projects Agency (ARPA) budget has actually been going up, and I will explain what is happening and how.

What I want to do is give a glimpse of what ARPA looks like today in terms of some of our investment strategies. These strategies do change over time; so I will try to point out some of the areas where things are changing. We have a set of technologies that we call core technologies. These are technologies we have invested in for some time; we expect to continue investment in these areas for as long as we can see in the future, and they are the basis for a lot of work that we do in actually developing specific military systems. One of the core technologies is information technology. Included are such programs as high-performance computing, the National Information Infrastructure (NII), research in software engineering, artificial intelligence, and communications technologies.

The second core technology is electronics technology, an area where we have also made investments for some time. It includes a wide range of technologies, from the development of sensors, such as infrared focal plane arrays, to the development of semiconductor manufacturing technologies. The SEMATECH program, which is jointly funded by ARPA and the semiconductor industry, is an example of electronics manufacturing. More recently, research has focused on three new challenges—packaging, flat-panel displays, and MEMS, or micro-electro-mechanical systems. The packaging program is developing manufacturing technologies for multichip modules, or MCM, in which several individual chips are interconnected in a single mechanical package. The flat panel display research is developing a variety of display technologies and their associated manufacturing technologies. The MEMS effort is using semiconductor manufacturing techniques to build arrays of very tiny mechanical moving parts.

And finally, there is the research in materials. We have worked for a long time in developing new materials, such as composites and artificial diamond—and the biggest problem now is being able to manufacture these materials at an affordable cost. It is one thing for the researchers to come up with a new material, it is now equally important to come up with manufacturing technology for that material. An example of a program where we have successfully developed the manufacturing technology is the infrared focal plane array program.

There is a broad area that we are now calling defense infrastructure, and it represents a new focus within ARPA that is being driven primarily by affordability of military systems. With the downsized defense budget, if you cannot design and manufacture a product at an affordable cost, you are really in trouble. The DOD needs to be able to manufacture low-volume semiconductor products at an affordable cost; we do not have the volumes of DRAMs or microprocessors. We need to be able to develop ships without having to create a mock-up of every piece before you build the ship. We should be able to reduce the acquisition costs of aircraft by up to 50 percent. These problems cannot be solved by technology alone; we also need to reform the defense acquisition system, and I will say more about that shortly.

Also, as part of the infrastructure are the projects in education and training, and these are also being driven by cost. If you look at the cost of training troops, which includes coordinating maneuvers involving armor, aircraft, and ships, there has been a major shift toward using distributed interactive simulation. Some of the programs we are working on right now make it possible to integrate what is done in the simulation world with what's being done in the actual test ranges. At some point, you may really not know whether the person you are talking to over the intercom is driving a real tank, or is driving a simulator, or the person who is dropping a bomb is dropping it at Nellis Air Force Base and you're seeing the simulated effects at the National Training Center. It is really very exciting because it brings in research in high-performance computing, communications, and virtual reality. It is a stimulating and challenging area that I believe will have applicability beyond DOD.

Health care is a new investment area for ARPA. We thought long and hard about it and decided to establish a five-year program, drawing extensively on what we have done in information technology and electronics technology, and applying these technologies to some of the problems in health care. In particular, we are looking at the problems of trauma care, where you first want to determine whether someone is injured and the extent of the injury, even though they may not be in your local vicinity; and second, to be able to do something about it. And finally, we are looking at the health information infrastructure. The DOD maintains a large number of hospitals in addition to the care that must be provided on the battlefield, and the cost of medical care for the DOD is eating a significant part of our budget. Here again we are being driven by affordability.

And finally, as you should expect, we have investments in "military systems." One area of investment is command and control, which draws extensively on our research in information technology and communications technology. You basically want to be able to think and plan inside the enemy's decision cycle, and to communicate your decisions to the forces. Our research in command and control is just as applicable to the civilian sector, if you think of all the things that go on in crisis management. Think of what happened with the recent earthquakes, fires, and hurricanes this country has experienced and the problems of not being able to get information to the people who need it.

A second area is combat vehicles. Aircraft procurement is one of the most costly elements in the defense budget, and ARPA has had a series of programs in aircraft development. We will continue to invest in these programs. We also have a program in simulation-based design for ships; and a new program called Maritech, which is modeled after SEMATECH, is trying to help the ship-building industry.

In our Precision Strike investment area, a major program called War Breaker focuses on detecting and targeting critical mobile targets. It does not help much if you find them two days later from aerial photographs or satellite photographs.

New areas we are exploring include counterproliferation and operations other than war. Even with the dissolution of the Former Soviet Union, we still have a major concern about weapons of mass destruction, whether they be nuclear, chemical, or biological. We have moved from a stable (though threatening) environment to one where there is a great deal of uncertainty from the military view point. Many of our current weapon systems do not help much in some of the new missions such as peacekeeping. The research challenge is to continue to provide our forces with the equipment for the missions they will encounter.

Let me talk about some of the characteristics of ARPA, particularly as they apply to academic research. First, we are a projects agency; it's in the name. And for us, it gives us great flexibility; it means that we can start and stop projects. If a new opportunity comes up, we can move in that direction. As an agency, we do not have laboratories or infrastructure that must be maintained, and there are no entitlement programs, although sometimes when you look at the way things are added to our budget by Congress, it almost feels like an entitlement program. The flexibility that we have and the fact that we can create projects, change the office structure if we need to, put critical mass onto a problem domain, really lets us do a lot. But we also terminate programs when we have completed what we set out to do. This can be a problem for universities, because the time constraint under which we operate may not be the same as that of universities, particularly if they are supporting graduate students.

We are part of DOD, as you know, and all the decisions we make are driven in part by what it means to the DOD. John Armstrong (in this volume) talked about the difficulty of investing in purely commercial technologies. We understand that very well. Even as we change our name from "DARPA" to "ARPA" and take on a dual-use responsibility, one of the uses is always the military use; if it does not have a military requirement, we should not be doing it. I will discuss later what dual use means for us.

A characteristic for our budget is that we fund everything from basic research (6.1 in the DOD nomenclature) through exploratory development (or 6.2), and on into the advanced development, which is 6.3a. The basic research program that we have is really the smallest part of our budget. We spend less than $100 million a year in basic research; the exploratory development has about $800 million. The rest of it, and the growth of the program, has all been in advanced development, and I will show in a moment what has caused this growth. The boundaries between different funding categories is not rigid, and we have sufficient flexibility to fund a wide range of research.

Let me now talk about some of the very broad trends at ARPA. Information technology is now the key technology at ARPA. Two of our nine offices are devoted to nothing but information technology, and every office in the agency is either a developer or a major user of information technology. So, this has been a change over time.

The second change is the role that manufacturing now plays in the agency. If you add up all the programs that have a manufacturing flavor, it is close to a billion dollars. A few years ago ARPA had an office called the Defense Manufacturing Office. We no longer have a single office called manufacturing, but rather, a number of the offices, probably at least half of them in the agency, are doing some form of manufacturing research. This is being driven largely by questions of affordability.

There are two new program areas at ARPA: health care, which I have already mentioned, and environmental research. In the environmental area, one thrust is in environmentally conscious manufacturing, with the current emphasis being on semiconductor manufacturing. We are beginning a new program in "green" manufacturing, where the complete life cycle of the product is taken into account, including the recycling or take-back of the product after its useful life. And finally, there is the problem of disposing of various kinds of waste material. ARPA has a supercritical water oxidation program to treat a variety of materials, including chemical warfare agents, propellents, and other DOD hazardous wastes.

Let me now discuss some of the trends that are affecting what we do.

First, let me talk about the "dual use" trend. Historically, about 70 percent of the ARPA budget has been spent on dual-use technology, so the name change from DARPA to ARPA and the greater emphasis on dual-use technology has not resulted in a significant change in the way ARPA operates. For many technologies, such as information technology, the most effective way to insert this technology into DOD systems was to make it available commercially. There have been a couple changes in the way we operate: We make a more conscious effort to look at both the defense and the commercial potential of our investments, and we are becoming more conscious of some of the business decisions that will affect commercialization of a technology or product.

Downsizing has had a major impact on defense acquisition and on the force structure, but it has not resulted in a decrease in the science and technology budget. There has been an indirect effect, though, since the downsizing in defense has made affordability a focus for much of our research.

The next trend, that of congressional earmarking, is a serious problem that has gotten worse over the past few years. Earmarking occurs when the appropriation committees write report language that "earmarks" funds for specific projects or organizations. The problem is actually broader than just earmarking. For FY94 nearly 70 percent of our budget has some form of congressional restrictions—earmarks, fences, or reprogramming restrictions. The Technology Reinvestment Project (TRP) had several earmarks attached to it; the TRP also had legislation requiring that it be competitive. The FY93 TRP selections have all been made, and all of the selections were made on the basis of merit. In some instances earmarked projects competed and won. In other cases earmarked projects are being funded with non-TRP DOD funds.

Another trend is toward more "interagency collaboration." For a long time ARPA and the National Science Foundation (NSF) have had a close working relationship and have done many projects jointly. We have now moved beyond bilateral collaborations and have established several multiagency projects. I will cite three. The Technology Reinvestment Project is headed by ARPA with participation by NSF, the Department of Energy, the National Institute of Standards and Technology, NASA, and the Department of Transportation.

The High Performance Computing and Communications program is a collaboration among ten agencies; each has money in its own budget, but there are cross-cutting activities that are coordinated by the Office of Management and Budget and the Office of Science and Technology Policy. Another example is the National Information Infrastructure in which there is extensive cooperation across most federal agencies. I personally feel that the cooperation that we see today is as good as I have ever seen it in the federal government across the agencies.

There is a trend toward doing more projects by consortia. The consortia frequently involve both industry and universities. I think it is a positive trend to see universities and industry working together. The TRP program, for example, required that consortia be formed. But it was happening even before the TRP came around, and I think it has been aided by the newly authorized funding methods. In addition to grants and contracts, ARPA is now using new funding vehicles called "agreements" and "other transactions" very extensively. This allows us much greater flexibility in the kind of agreements that we strike between the government and the contractors. For example, we have frequently renegotiated the intellectual property rights to provide a greater incentive to industry to commercialize the results of their research.

Finally, there is defense conversion. I've already mentioned the Technology Reinvestment Project (TRP). For the TRP and other efforts in defense conversion to be successful, they must be accompanied by acquisition reform. For DOD, in particular, as we downsize and as the procurement budget decreases, the corresponding overhead caused by the accounting requirements, military specifications (MILSPECS), extensive documentation, and various things that have driven the cost up have not come down proportionately. So, the DOD's real procurement budget is getting squeezed by more than just the downsizing, and only real acquisition reform will be able to change this trend.

Let me now discuss the ARPA budget. Figure 1 is a plot of the ARPA budget in current dollars; it has not been adjusted for inflation. Notice that starting in FY93 the budget ramps up fairly significantly and then holds constant at about $2.5 billion through fiscal year 1999. I expect it is actually going to go higher.

Figure 2 shows the budget for three fiscal years, 1993, 1994, and 1995. In FY93 the ARPA component of the President's budget was $1.3 billion. We recently tallied up how much money we managed during FY93, and it was over $2.7 billion dollars. And, by the way, that was with no additional people in the agency. We only have about 200 people; I think we are about one-sixth the size of NSF, and our budgets are approximately the same.

Here's what happened: There were some large programs in the Pentagon, one called the air defense initiative, and one called the balance technology initiative, worth combined about $300 million, which were managed

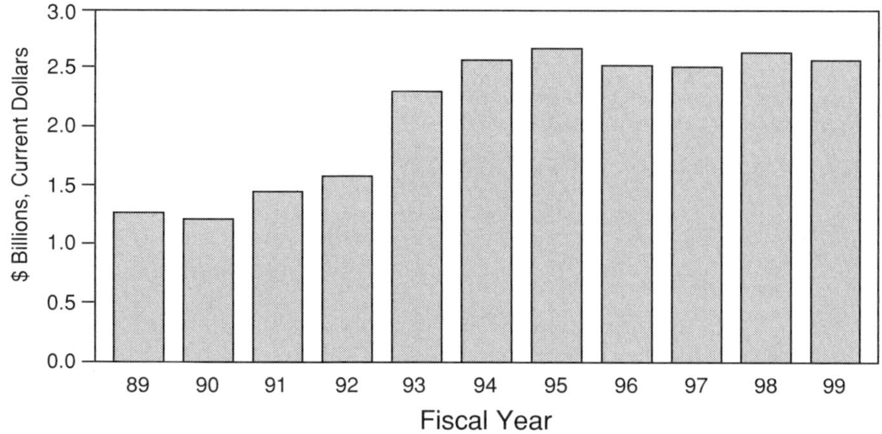

FIGURE 1 ARPA budget, actual and estimated, in current dollars. SOURCE: Advanced Research Projects Agency.

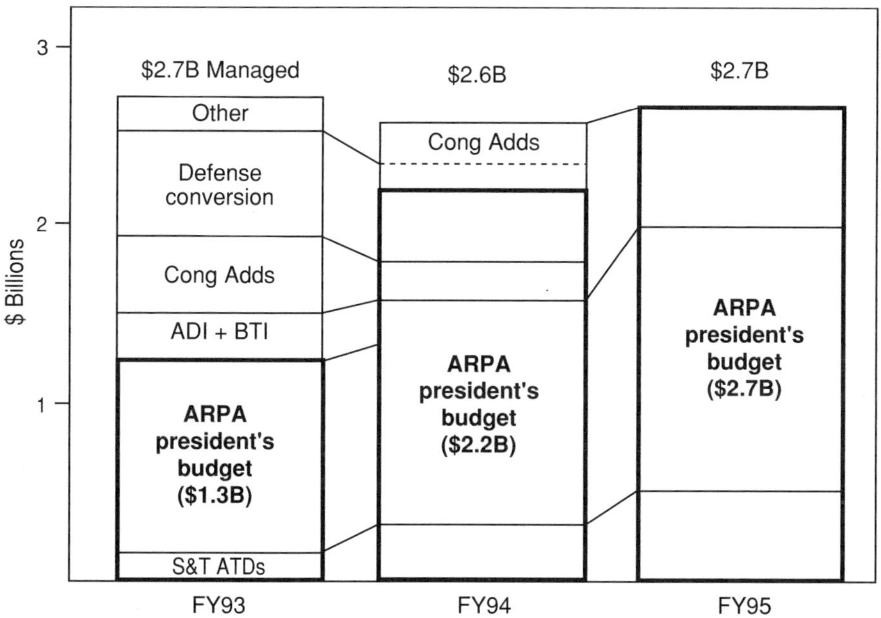

FIGURE 2 ARPA budget FY 1993 to FY 1995. SOURCE: Advanced Research Projects Agency.

by ARPA. In FY94 they became part of the ARPA budget. When our FY93 budget was submitted last year, Congress added almost a billion dollars to it. More than $500 million was added for "defense conversion"; it is what became the TRP program. And then there were a number of programs that Congress had been funding on a year-by-year basis: Examples include flat-panel display technology, lithography, electronic packaging, and some research in materials. It is very difficult to run a program if you get money on a year-by-year basis without any out-year budgeting, and so you really have to hedge your bets with a lot of options or you forward fund an entire program.

As part of the President's FY94 budget, the Defense Department included about $325 million for the TRP. A number of the congressional programs were also funded as part of the basic defense budget for ARPA, and the budget grew to about $2.2 billion. The final budget that was appropriated by Congress contained $475 million for defense conversion and a number of other programs added by Congress, raising the total to more than $2.6 billion. A number of our proposed programs were not funded.

For the out-years, from fiscal 1995 to fiscal 1999, ARPA will have more than $600 million a year in the TRP. That is almost a quarter of our budget for this program. We cannot predict what Congress will do when the budget is submitted, but I suspect that the budget will be pushed upward toward the three billion mark.

Figure 3 shows how the universities have fared with ARPA funding over the years. These figures are, again, in current dollars and you can see that there has been an increase. In FY93, $300 million went to universities. It is getting harder to do the accounting because of the TRP program, and I will explain that in just a minute.

Figure 4 shows the university funding as a percent of the total ARPA budget, and you'll notice that it has more or less leveled off at about 17 percent of our total budget. It is down in FY93 because of the TRP. Figure 5 shows what is happening. First, its shows how the funding varies by the various ARPA offices. The Computing Systems Technology Office (CSTO) and the Software and Intelligent Systems Technology Office (SISTO) account for nearly 45 percent of the total university funding. The second point deals with the TRP. On the order of $60 million is going to universities. Part of that is for the manufacturing education and training component of the TRP, and part of it reflects universities as members of a number of TRP consortia. At this time we do not have an adequate way to project the amount of consortia funding that goes to universities or to any particular company.

There is one additional point to make about university participation in the TRP. Figure 6 shows that nearly 40 percent of the development proposals that were submitted contained at least one university participant and

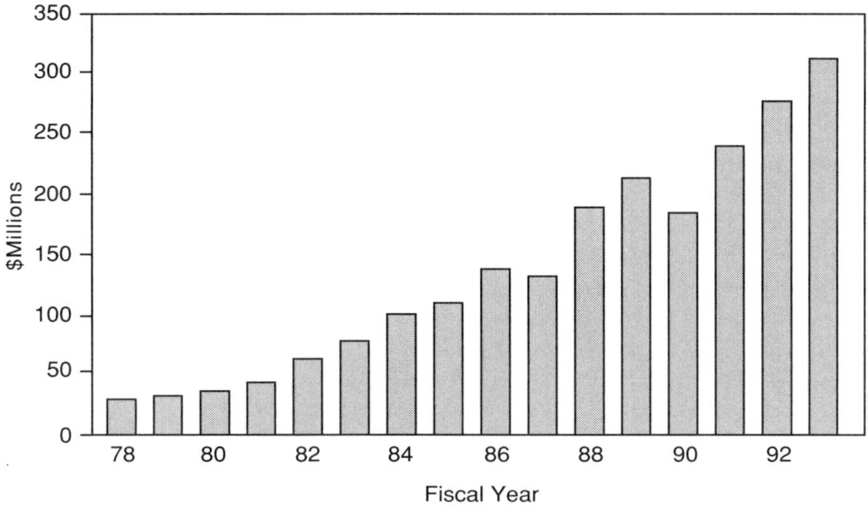

FIGURE 3 ARPA funding to universities in current dollars, 1978 to 1993. SOURCE: Advanced Research Projects Agency.

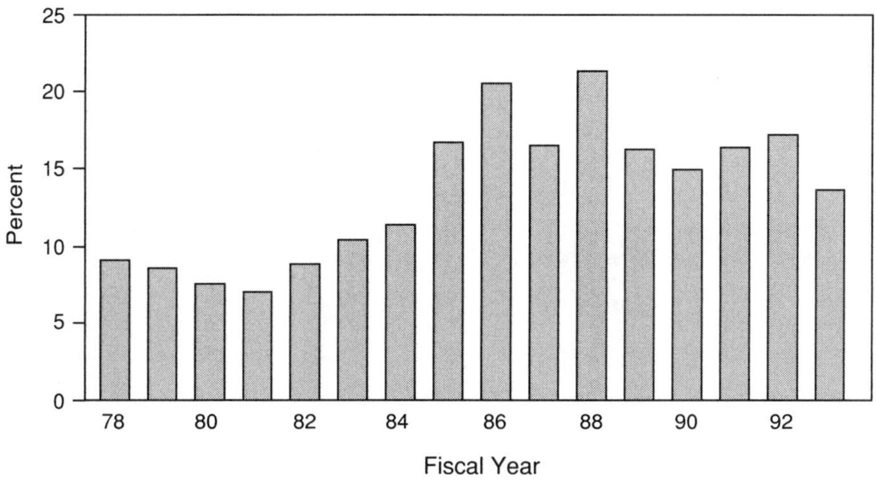

FIGURE 4 Percent ARPA budget allocated to universities, 1978 to 1993. SOURCE: Advanced Research Projects Agency.

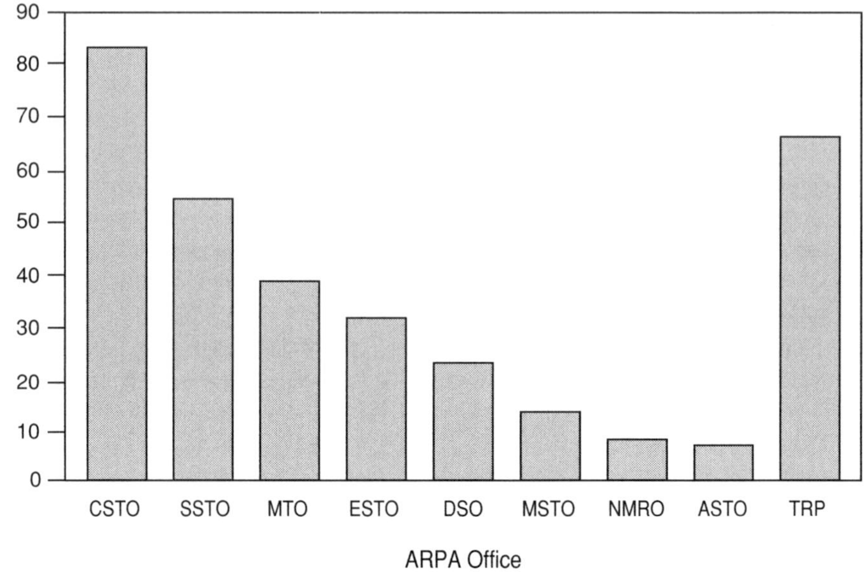

FIGURE 5 FY 1993 funding to universities, by ARPA office. SOURCE: Advanced Research Projects Agency.

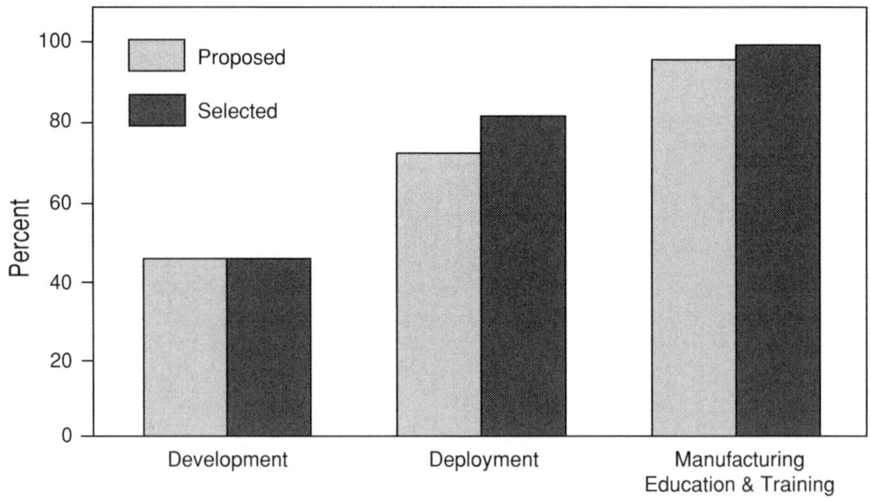

FIGURE 6 Percent of TRP proposals and funded projects with at least one academic institution participating. SOURCE: Advanced Research Projects Agency.

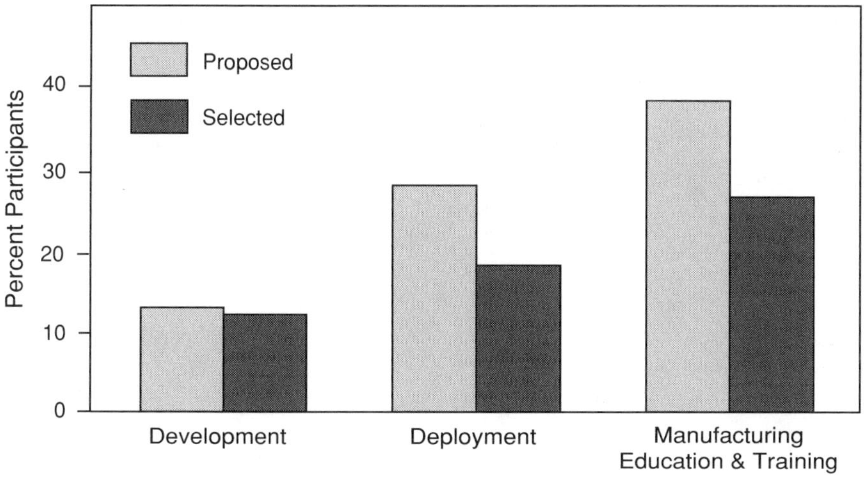

FIGURE 7 Percent university participation in TRP proposals and funded projects, by ARPA category. SOURCE: Advanced Research Projects Agency.

nearly 100 percent of the proposals in manufacturing education and training contained at least one university. If you look at the total number of organizations that participated in the TRP—universities, companies, national labs—you will find that over 10 percent of the development participants and nearly 40 percent of the manufacturing education and training participants were universities. (See Figure 7.)

Let me turn now to some of the opportunities for the university research community. The first is to develop better collaboration with industry. I think it is absolutely essential for both the universities and for industry that we have this collaboration. One of the mechanisms that I see making this happen is the funding mechanisms called agreements and other transactions. This allows us to structure some very flexible arrangements that do not force either the university or the industry to be a subcontractor to the other.

I think the collaboration with industry is beneficial to the universities because it exposes them to some of the real problems that we in defense want to have solved. For industry, I think the earlier they understand what is going on in some of the new technologies, the easier it will be to incorporate these technologies into their projects. In some cases, such as in data storage, the universities are playing a major role in developing the technology; they are almost the development labs for industry in some instances.

The next three opportunities are closely related. The first is the notion of interdisciplinary, or multidisciplinary, research. For many of the most difficult problems that we need to deal with, such as in robotics, manufacturing, or data storage, this collaboration—this interdisciplinary activity—is absolutely essential. Today's tough problems do not map into the traditional disciplines around which most university programs are organized.

The second of the related topics is systems research. Systems are often hard to deal with in a university environment because of their size and complexity. Yet, there are fundamental problems that we do not understand. We do not understand much about how large systems scale, and the systems we build are going to be increasingly software intensive. This is an area in which collaboration with industry may be very appropriate.

Finally, there is the question of applications and the role of scientists or engineers in helping develop some of the applications. Much of the applications research in the high-performance computing program has focused around applications called "grand challenge" problems. These are problems that are very hard, often pushing the edge of our understanding and the edge of our technology.

Besides the grand challenge problems, which I think have extensive university involvement, there are what we are calling the "national challenge" problems. These are problems that have a broad and direct impact on the nation's competitiveness and the well-being of its citizens. These problems include manufacturing, health care, and education, among others. I think it is important to begin to focus the technologies that we are developing on some of these problem domains.

The last topic I would like to address is personnel exchange at ARPA. We do not have any people to send back to universities, but many of our program managers come from universities under the Intergovernmental Personnel Act (IPA). We currently have about 20 IPAs and we can hire another 15. This is an opportunity for researchers from the university community to come and work with the government for a period of two to four years. Two years is the absolute minimum; four years is desirable. If you know of people who are interested in this, send them to see us, or encourage them to take a sabbatical and spend some time working with the government. NSF offers the same opportunity. I think it would be beneficial to the individual; and it would certainly help us manage our programs and keep abreast of the technologies.

What Should Industry Expect from Academic Engineering Research?

John A. Armstrong

Before turning to what I feel are the central issues confronting academic engineering research today, I want to give a list of topics that are not, in my view, the most important issues. Matters of indirect cost recovery, of getting sponsors to pay the full costs of research, matters of fraying university research infrastructure are all important, but they concern a large population of university departments, so I do not think they are peculiar to engineering research. Nor do I think it is very important to concern ourselves about undue influence by industry on the academic research agenda. There is no way to have a determining effect on that agenda without putting up substantial amounts of money, and I do not think that industry is likely to do so in most fields. On the other hand, mutually beneficial relationships between engineering research departments and their related industrial sectors are matters of concern, and I will have something to say on that topic later in this paper.

In addition, although the changes in corporate R&D investments are clearly relevant to the future of academic engineering research, I believe it is misleading to describe these changes as the "apparent abandonment of in-house capacity for basic industrial research." There is more to it than that, and I will have more to say about that later in this paper as well.

Finally, I do not think that we should be talking about "the cures for national competitiveness concerns being found in academic engineering departments" if that is taken to imply the simplistic notion that competitiveness problems stem from lack of technology transfer of new results from academia to the private sector. The competitiveness problems of American industry have almost nothing to do with lack of technology transfer from universities, or national labs for that matter. The responsibility for deficiencies in our industrial performance rests largely with failures in the private sector, failures of strategy, investment, and training—failures, in short, of management. There have also been failures of government policies in the areas of macroeconomics, access to overseas markets, and the balance between consumption and savings. Neither the private nor the government failings will be cured, or even helped, by more engineering research. Trying to cure poor industrial performance in the short term by more university research is like asking for helpers when pushing on a rope.

Having said that technology transfer is not one of the real issues, I hasten to say that I do believe that academic engineering research can make a major contribution to improving the ability of our nation to realize the fruits of its investments in basic and applied research. But that contribution will have more to do with the nature of the advanced training given to Ph.D.'s than with the specific research results produced as part of that training.

I want now to summarize what I think are a few of the real issues confronting academic engineering research today, as seen from the industrial perspective.

Recently I gave a Compton Lecture at MIT titled "Is Basic Research a Luxury Our Society Can No Longer Afford?" My answer was "No, it is not a luxury, it is a source of national competitive advantage." But that answer was strongly qualified. I argued that national leadership in basic research was neither necessary nor sufficient for society to achieve its economic and environmental goals. The reason is that successful R&D represents less than 5 percent of the process by which wealth and jobs are created. Countries that do much or all of the other 95 percent in a world-class way can be successful in reaching their goals without being world leaders in research. But I went on to assert that a nation that does the 95 percent competitively, *and* leads in basic research, may expect to have a comparative advantage. And since the United States currently enjoys world leadership in many areas of research, we ought to be careful to preserve that advantage while, in parallel, we address deficiencies in our national performance in the 95 percent of wealth creation that is not R&D.

It is in thinking about the larger role that scientists and engineers can play in some of these "downstream" activities that I have been led to ask the question "What is an Engineering Science Ph.D. For?" In what ways can academic engineering research contribute more effectively to the rest of the processes by which new knowledge is turned into societal value, processes that for the most part lie outside traditional science and engineering, but to which, experience has shown, engineers and scientists can make major contributions? In my view, this is the overriding issue confronting academic engineering research today.

But before expanding at length on this question, it is important to keep it in perspective with a set of other important considerations. The first of these additional considerations is that the possible role of the government vis-à-vis research in support of civilian technology is very different from the government role (which has been eminently successful) in supporting military technology. The government was the customer for military systems, and as such could guide and control its support of research by whether its clearly understood needs were being met, or were likely to be met. But the government is not the definitive (nor even a very important customer) for most civilian technologies. (Indeed the government is a customer that drives most companies "up the wall" because of its cumbersome, often wrong-headed, and counterproductive procurement practices.) It is not expert on the results to be achieved with civilian technology nor on how to measure success. Therefore it should exercise a healthy degree of caution in creating policies and programs to direct research funds toward nonmilitary national goals and in pushing for technology transfer from the national labs to the civilian sector. I do not see sufficient appreciation of these dangers in the public discussion, although the plan to channel funds to universities through the industrial members of industry-university consortia is clearly an attempt to deal with this issue. Whether that will turn out to be a good idea or not, time will tell.

The next consideration to be kept in mind relates to changing industrial R&D portfolios. While many industries have been reassessing what types of applied and engineering research are likely to be of significant help to them in achieving comparative advantage in the coming decade, the engineering research portfolios of universities will probably change much too slowly to be in step with the needs of many sectors. This lack of being in step is more a matter of balance than a lack of appropriate investments altogether. Said another way, academic engineering research is well adapted to creating new programs (witness the rapid emergence of programs relating to manufacturing). But academic engineering research (and all other forms of academic research as well) are just about incapable of stopping programs, of scaling back investments, and of redirecting the work of faculty. This guarantees a number of glaring mismatches between academic engineering research and industry in the coming decade. Clearly I have a different view than many of my academic colleagues about the problems of undue industrial influence!

So, before we conclude that downsizing and other changes in corporate R&D portfolios are all regrettable and shortsighted, we should ask whether particular R&D areas are still such good investments as they may once have been. Not only does the leverage of a particular field of research in industry change with time, but there can also be a change in the relative importance of the university engineering research contribution to such a field.

Of course, this balance has shifted more in some fields than in others. The newer the field, and the less the aggregated industrial R&D investment, the greater the significance of university-derived results.

Whereas today's best example is the whole set of technologies spawned by progress in understanding the molecular basis of life, in the 1960s and 1970s electronics and computer science were the leading examples. Then, the relative contributions of university-based research results in electronics and computer science were much greater than they are today. But now the resources devoted to R&D in industry, and the accumulated knowledge, know-how, and investment, make industry less dependent than was the case twenty years ago.

Ironically, during the 1980s there was substantial attention to building silicon-technology-based research facilities in universities just before it became clear to the electronics and computer industries in the 1990s that the value of hardware technology and expertise was declining in relation to that of software, systems, and applications expertise. Industry's response in some cases was to leave hardware technology development altogether, and in other cases to scale back the investments being made, often by forming consortia to share risks and investments. What academia's response to these developments will be is far from clear.

How might university engineering research have avoided "zigging" just as the electronics industry "zagged?" Better relations between industry and academia, focused more strongly on their principal enduring common interest, might have helped. That "principle enduring common interest" is, of course, highly trained students.

It follows, I believe, from all of the considerations I have listed that academic engineering research may need to rethink the importance it attaches to research results per se in relation to the value of the Ph.D. training through which those results are obtained. In a word, for the next decade or so, the training will be more important than the research results, at least in many fields. This is an institutional issue as well as an issue to be faced by university faculty as individuals. Clearly it is an issue for the mission agencies that support academic engineering research as well. They need to reassert that they have an explicit mission to foster graduate technical education as well as to support research whose results are useful to them.

In any future rethinking of engineering science Ph.D. programs, one should examine not only the appropriate portfolio of technical areas and programs maintained, but also ask, How can academic engineering research be more effective in helping the nation achieve its goals for more and better jobs, a rising standard of living, and a more sustainable relationship with the environment?

In short, one should ask the question, "What is an engineering research Ph.D. for?" Although I am only a visitor to academia, I propose to devote the balance of my remarks to addressing this mildly provocative question.

Many engineering schools have recently reassessed their master's degree programs, and many schools have made significant changes in those programs. It is also true that many professional schools have thoroughly revamped their curricula, expectations, and culture. A good example of this is the New Pathway program at the Harvard Medical School. But there has been little serious reassessment so far of the underlying assumptions, expectations, and requirements of Ph.D. programs in science and in areas of engineering closely allied to science, the areas we have been calling engineering research. In my view it may be time for such a reassessment.

Of course, in many respects the Ph.D. programs in science and engineering are in good shape. The technical sophistication of new graduates in their specialties is often breathtaking. New Ph.D. graduates are still the best "vehicles" in the world for transfer of new insights and new ways of doing things.

And yet . . . there are serious problems as well, problems I came to see over many years of hiring and managing new Ph.D.'s. In brief, it is my view that the training of new Ph.D.'s is too narrow intellectually, too campus-centered, and too long. Furthermore in my experience, many new Ph.D.'s have much too narrow a set of personal and career expectations. Most do not know what it is they know. They think that what they know is how to solve certain highly technical and specialized problems, like designing microprocessors or writing high-speed networking protocols.

Of course, what they actually know that is of lasting value is how to approach and solve problems starting from powerful and fundamental points of view. But to my surprise, most do not understand that that is what gives them any edge they may have over young people of their own age who are already out in the workplace without a Ph.D. but with a six- to eight-year head start in experience.

This is all part of what one might call the Ph.D. paradox. The phrase is simply a way of drawing special attention to what we all know, but which is not, I think, sufficiently taken into account in the design of Ph.D. programs. To earn a Ph.D. in engineering research, a young person is expected to make an original contribution to fundamental engineering science. To get to the frontier, it is expected that one will ask a narrowly defined set of questions, and in that narrow region, think or experiment deeply. In the course of this deep but narrow exploration, the graduate student acquires a powerful methodology for formulating and solving technical problems, starting with an understanding of the fundamentals of the subject. He or she learns how to pose a problem, decide what data or experiments are required to solve it, obtain that data, analyze it critically, and then defend the conclusions vigorously. He or she has learned how to acquire new skills, including the ability to understand and use just about any form of applied mathematics. The Ph.D. candidate has, in a word, learned how to learn at a very sophisticated level.

The "paradox," of course, is that in the course of deep, specialized inquiry, one acquires an intellectual armamentarium and outlook of great general utility. The training of the scientific or engineering specialist in fact provides much of what might be termed training for the advanced technical generalist. It is a further part of the paradox that many new graduates do not seem to value this powerful generalist capability—perhaps because their professors do not value it either.

This overspecialization often has unfortunate consequences for new engineering scientists. Overspecialization can result in a lack both of perspective and of self-confidence; new Ph.D.'s often believe themselves ill prepared to venture outside their specialty to use their powerful training in jobs in development, manufacturing, and technical management, let alone in tasks even farther afield from their specific training. The burden of overspecialization is compounded by their often total lack of work experience outside the university and by a culture that often suggests to them in not so subtle ways that becoming like their professor should be their goal and mark of success.

This paradoxical situation is due in part to the lack of serious requirements for scientific and technical breadth in the typical graduate curriculum, as well as to the fact that there is little or no encouragement, and a lot of implicit discouragement, for the young person who wants to spend time during graduate school off campus in a setting where technical knowledge is actually used. There is, in short, almost no value assigned to technical breadth or to real-world experience as an essential part of Ph.D. training.

You may recall that I asserted that the typical Ph.D. degree takes too long to acquire. I firmly believe that to be the case, and I see no contradiction between shortening the time to obtain a Ph.D. and my just expressed desire to see young people spend more of their time away from campus as part of their training.

I hold these seemingly contradictory views because of a hypothesis I have about why the typical Ph.D. takes so long. It is only in part because of course requirements and faculty pressure to get more research results for a thesis. It is due in large part both to the students' comfort with graduate student life and to their anxiety about what it will be like in the outside world when they leave the university. This is all possible, of course, because universities and funding agencies permit and support such long stays.

If I were a sociologist I would test the following hypothesis, both retrospectively and prospectively. What is the average length of time to the Ph.D. of young men who are married and have small children while they are graduate students? The answer I expect is that it is up to two years shorter than the average. (That the opposite result will obtain for young married women graduate students is altogether a different problem!)

Just as experience of family responsibility tends to shorten one's tolerance for the life of a graduate student, so, I believe, will experience out in the world of technical work tend to lower the typical graduate student's anxiety about finding a job and starting a career.

Now we cannot require graduate students to get married and start families, but we could exert serious pressure to bring the average duration down by a year or 18 months. Shortening the average duration of graduate study will lower the cost to the nation for training a given number of young scientists and engineers, and it will saddle the graduates with less of a disadvantage with respect to their contemporaries who are years ahead in gaining experience and seniority in the workplace.

What can industry do to help? It can and should be responsive to setting up cooperative arrangements with engineering research departments. I believe that small firms and start-ups have the most to gain by such arrangements, and also the most to give students in the way of broad perspective. Many of our best graduate schools are surrounded by such small companies, many of which have been started from university science and engineering programs. However, except for the students of faculty members connected with these spin-offs, these exciting firms are invisible to the majority of graduate students.

I know well, by the way, that it is hard, time-consuming work for faculty members (as well as for their industrial colleagues) to set up mutually advantageous joint projects involving graduate students. But because they will, on average, contribute so much to the improvement of the student's education, both faculty and industry managers should make the time.

There is more that industry can and should do. Companies should be more willing than they now are to have key technical people spend time in universities as adjunct faculty. The improved perspective they will bring to faculty and to graduate students will be more than enough to offset the substantial effort it will be to initiate such arrangements. Similarly, there is far from sufficient value placed on faculty members having professional experience in the outside world. And by that I mean more than the casual knowledge that consultants obtain of the culture, the problems, and the intellectual value that exists in off-campus engineering research.

It is true that, both as individuals and as members of their discipline, professors take pride in the fact that many of their students turn out to have highly successful careers in industrial management, or in government service, or in the business world generally, or as teachers and professors in nonresearch institutions. But this is all thought to be irrelevant to the graduate curriculum. The curriculum is still characterized overwhelmingly by what is necessary for the training of future research faculty members.

Although these nontraditional uses of the Ph.D. have been around for a long time, their importance both to society and to society's support for the scientific research enterprise requires that they be taken into account in new ways. The reason is that, as described in my first Compton Lecture, getting the R&D done "right" is less than 5 percent of the job of turning new knowledge into the social and economic utility for which society supports scientific research in the first place (including basic research). The other 95 percent of the job has to be done in a world-class, competitive way if the society that pays for the research is to be the society that gets a fair return on its investment.

In the doing of the other 95 percent of the job, many people with skills outside of science are needed, to be sure. But even so, there is much of the 95 percent of the job that is not R&D that can and should be done by people with scientific training. I say "can" because much of this work is done best by those with technical background and understanding; and I say "should" because societies that do bring technical generalists to bear on this work will have an advantage in world competition and will get more for their investment, sooner, than countries whose scientists and engineers play less frequent and prominent roles beyond the laboratory.

The presumption seems to have been that the apprenticeship process designed for the traditional science Ph.D. degree would do as well as needs to be done in fitting graduates for employment as science and engineering Ph.D.'s in what I have called the nontraditional roles. Certainly the traditional Ph.D. training is not bad preparation for nontraditional roles, but it is hard to believe it cannot be done better. Society's support for academic research may even depend on its being done better.

Indeed, I believe society is poised at this moment between developing an enlarged expectation of what scientists and engineers can do, on the one hand, and concluding that we have been largely overrated in our contributions to society, on the other hand. If this perception is correct, it behooves us all to take the improvement of graduate science and engineering education very seriously.

I have already said that industry can and should do more to help enlarge and augment graduate technical education. But it must also be said that those of us outside the university cannot possibly be major actors in this reassessment and revamping. The most that we in industry can do is offer to help where appropriate and to transmit our sense of the urgency of the task of rethinking Ph.D. training. We feel this urgency because the students at issue are of enormous importance to our own future and because we believe that society's continued support of the university engineering research establishment depends in no small part on that establishment's doing, and being seen to do, a better job at fitting technically trained citizens to play their full role in achieving the goals of society to which science and technology can contribute.

Background Paper

The Academic Engineering Research Enterprise: Status and Trends

Charles H. Dickens

The purpose of this paper is to describe the status of U.S. academic engineering research universities and several major trends affecting them. The paper provides summary descriptive text and an appendix with tabulations selected from a variety of data sources.

"Engineering research universities" are those institutions that reported research and development (R&D) expenditures for engineering or computer science in the 1991 National Science Foundation survey of R&D expenditures of universities and colleges. There are 219 such institutions.[1] For the purposes of this paper, the term "engineering" includes computer science.

The paper is divided into four major sections and a data appendix. The first section describes the recent history and current status of the U.S. academic engineering research structure. Some major characteristics of the 219 engineering research universities are presented, including activities of organized engineering research centers and laboratories and federal programs that support engineering research centers. Characteristics of faculty and other engineers employed by academic institutions are described. Postdoctoral fellows are discussed in terms of their distribution by field, gender, and sources of support. Information on student enrollments is presented for undergraduate and graduate students by field, gender, and minority status. Trends in bachelor's, master's, and doctor's degrees awarded are presented by field and gender.

The second major section presents funding of academic engineering research. Government sources of support for academic research by field and category of research are described. Trends in research and development expenditures at engineering research universities are presented by field and source of funds. Support sources for graduate students as research assistants and for postdoctoral appointees are described. Information is presented on the mission basis for government research support. Support for academic engineering research from industry, universities' own funds, and from foreign sources is described.

The third section addresses the nature and scope of relationships between engineering research and education. Topics covered include participation in research by undergraduate students, graduate engineering students, and engineering faculty, postdoctorates, and other academic engineers.

The fourth section includes definitions, limitations, and principal data sources used in this paper. Significant gaps in currently available data are discussed.

RECENT HISTORY AND CURRENT STATUS OF THE U.S. ACADEMIC RESEARCH STRUCTURE

This section describes U.S. engineering research universities, organized engineering research units, and major human resources for engineering, including faculty and other engineering employees, postdoctoral appointments, student enrollments, and degrees awarded.

Engineering Research Universities

In 1991 there were 219 universities and colleges that reported research and development expenditures for engineering and computer science. (See Table 1, Dickens Appendix.) Of these institutions, 168 reported R&D expenditures for both engineering and computer science, 37 reported expenditures for engineering only, and 18 reported expenditures for computer science only. The majority of the 219 institutions were public (158), and 61 were private.

Since these 219 institutions were selected on the basis of their R&D expenditures for engineering and computer science, they include a variety of universities and colleges when viewed in terms of other classification systems. For example, the 1994 Carnegie Classification for these 219 institutions is as follows: Research Universities I - 83; Research Universities II - 36; Doctoral Universities I - 30; Doctoral Universities II - 32; Master's Universities and Colleges I - 29; Baccalaureate Colleges I - 3; and Professional Schools and Specialized Institutions - 6. The Research I Universities accounted for 70 percent of the R&D expenditures for engineering and computer science in 1991.

The 1991 National Science Foundation survey of graduate enrollments reported 1,464 graduate engineering departments, of which 1,260, or 86 percent, were at doctorate-granting institutions. There were 274 computer science departments with graduate enrollments, with 187, or 68 percent, of these departments at doctorate-granting universities.

Engineering Research Centers and Laboratories

There is great variety in the internal organization of engineering research universities. In addition to departments, there are a large number of engineering research centers and laboratories, which may or may not be within departments or even within engineering colleges.

In an NSF-funded study currently under way, Robert P. Morgan and his colleagues identified 1,030 organized, university-based engineering research units at 154 universities within the study population; there may be others. These research units were defined "very broadly to include units that either are totally within engineering schools or that may not be within engineering schools but involve engineering faculty and staff."[2] Morgan and colleagues found that these organized research units were relatively recent organizations, with one-half being founded since 1983. Many of these units were created "to provide a focal point for certain research activities and to attract funding and facilities."[3]

The research activities of the units surveyed by Morgan and colleagues included a broad range of engineering disciplines. The overall distribution of research effort as described by the responding unit directors was about equally divided among basic research, applied research, and development. In addition, Morgan and coauthors reported that, when asked into which of six broad critical technology areas the work of the units fell, the directors indicated the following divisions:

Materials	45%
Energy and environment	42%
Manufacturing	29%
Information and communications	27%
Aeronautics and surface transportation	17%
Biotechnology and life sciences	13%[4]

Federally Sponsored University Center Programs. Six federal departments and independent agencies sponsor university research centers, many of which have an engineering focus. A 1993 report of the National Research Council's Transportation Research Board reported 281 centers being funded through nine federal programs:[5]

	No. of Centers
U.S. Department of Transportation	
University Transportation Centers Program	13
National Science Foundation	
Engineering Research Centers Program	18
Science and Technology Centers Program	25
Materials Research Laboratories	10
Industry-University Cooperative Research Centers Program	50
National Institute of Standards and Technology (NIST)	
Manufacturing Technology Centers Program	7
National Aeronautics and Space Administration (NASA)	
University Space Engineering Research Centers	8
Department of Defense	
University Research Initiative	113
Department of the Interior, Bureau of Mines	
Mineral Institute Program	37

NSF-Funded Engineering Research Centers. In 1985 the National Science Foundation established the Engineering Research Centers (ERC) Program in accordance with a model envisioned by the National Academy of Engineering. The program was motivated by three major concerns: To restore U.S. industrial prowess in turning research discoveries into high-quality, competitive products; to give greater emphasis to the design of manufacturing processes and products; and to better prepare engineering graduates to meet the needs of U.S. industry. Each ERC is established as a three-way partnership involving academia, industry, and the National Science Foundation. Annual funding for an ERC ranges from $2.5 million to $8.0 million, with the NSF contribution ranging from $1.8 million to $3.3 million a year. The fiscal year 1995 budget requests $51.5 million for the ERC program. The distribution of the 18 current NSF ERCs by major technological area of focus is as follows:[6]

Design and manufacturing	5
Materials processing for manufacturing	3
Optoelectronics/microelectronics/telecommunications	4
Biotechnology/bioengineering	3
Energy and resource recovery	2
Infrastructure	1

Faculty and Other Engineers Employed by Academic Institutions

The engineering R&D activities of research universities rely heavily on faculty, nonfaculty research staff, postdoctoral appointees, and graduate research assistants.

Doctorate-holders employed by academic institutions. Over the period 1979 to 1989, the overall employment of doctoral engineers and computer specialists increased by 72 percent. (See Table 2, Dickens Appendix, for data on academic employment of doctorates.) All fields experienced growth, ranging from 23 percent for materials engineering to 178 percent for computer science. The large percentage increase for computer specialists reflects the small number in the base year. The proportion of doctoral engineers and computer specialists who were active in research and development increased from 76 percent in 1979 to 79 percent in 1989. There were variations by field in the staff active in research and development. Increases were noted except for aerospace and civil engineers and computer specialists. The proportion of chemical engineers active in research and development had the largest gain, increasing from 73 percent in 1979 to 92 percent in 1989.

Faculty and nonfaculty research staff. Compared with other aspects of the academic engineering research enterprise, there is sparse information on faculty in universities and colleges.[7] In academic year 1992-93, there were more than 21,000 engineering faculty at U.S. universities and colleges.[8]

The U.S. Department of Education, with the cosponsorship of the National Science Foundation, is conducting the "1993 National Study of Postsecondary Faculty." When completed, this study should provide substantially more information than has been available on the characteristics and activities of faculty in engineering, computer science, and other fields. The coverage of separate engineering fields, however, is limited to the following: general engineering; civil engineering; electrical, electronics, and communications engineering; mechanical engineering; chemical engineering; other engineering; and engineering-related technologies.

According to a similar, but less-detailed, U.S. Department of Education survey for academic year 1987-88, the full-time regular instructional engineering faculty (including engineering-related technologies) in postsecondary education were predominately male (98 percent) and predominately white, non-Hispanic (87 percent). The majority (64 percent) of the engineering faculty held doctorates. The distributions of engineering faculty by age and academic rank were somewhat like those for the natural sciences, except that a higher proportion of engineers were in the oldest category and there was a higher proportion of assistant professors in engineering.[9]

	Age		*Academic rank*		
	Less than 40	60 and older	Prof.	Assoc. prof.	Asst. prof.
Engineering	23%	14%	41%	24%	23%
Natural sciences	23%	9%	38%	23%	18%

The numbers of engineering faculty have increased over the years since a 1980 NSF-sponsored survey found that there were 16,200 permanent, full-time engineering faculty positions.[10] According to a 1986 National Science Foundation survey of doctorate-level departments in six engineering fields—aeronautical, chemical, civil, electrical, industrial, and mechanical—there were approximately 9,800 full-time faculty in these departments. About 70 percent of these faculty were tenured. These departments also reported 615 nonfaculty doctoral personnel who were employed full-time as professional researchers.[11]

The engineering departments surveyed in 1986 by the National Science Foundation reported that the full-time faculty had submitted more than 14,200 research proposals during the previous year (defined as July 1, 1984, to June 30, 1985). In contrast, members of the nonfaculty doctoral research staff were much less likely than faculty members to submit research proposals on which they would be the principal investigator; for this group the number of proposals submitted was about 240.[12]

Postdoctorates. Postdoctoral fellows and associates form a substantial part of the research staff at doctorate-granting academic institutions. (See Table 3, Dickens Appendix, for data on postdoctorates by field, citizenship, and gender.) In 1991 there were 2,406 postdoctoral appointees in engineering and computer science departments, almost all of whom (2,394 or 99.5 percent) were at doctorate-granting universities. Over two-thirds (68 percent) of the postdoctorates were in four fields:

Chemical engineering	25%
Materials engineering	17%
Mechanical engineering	14%
Electrical engineering	13%

Non-U.S. citizens held the majority of postdoctoral appointments in all fields of engineering. The overall proportion of non-U.S. citizens in 1991 was 70 percent; by field, this proportion ranged from 30 percent for computer science to 80 percent for materials engineering.

The number of postdoctoral appointees in engineering and computer science departments grew dramatically between 1980 and 1991, increasing 136 percent. By field, the increases ranged from 52 percent in civil engineering to 285 percent in aerospace engineering. Chemical engineering, which had the largest number of postdoctoral appointees in 1991, had an increase of 215 percent over this period.

Women increased their overall representation among postdoctoral appointees in engineering and computer science from 7 percent in 1980 to 11 percent in 1991. Chemical engineering had the largest share of the female postdoctorates in 1991, 35 percent. (Chemical engineering had 23 percent of the male postdoctorates in 1991. See Table 4, Dickens Appendix, for data on postdoctoral appointees by field and source of support.) The number of female postdoctorates in computer science departments grew from 2 in 1980 to 27 in 1991, but the large variations in their numbers over this period made it difficult to give a precise sense of their share of the total. In 1990, for example, 13 percent of the computer science postdoctorates were women, compared with 18 percent in 1989 and 17 percent in 1991.

Overall, federal sources provided the support for two-thirds (67 percent) of the postdoctoral appointees in engineering and computer science departments at doctorate-granting universities in 1991. The principal mechanism of federal support was through research grants, which accounted for 94 percent of the federally supported postdoctorates. Except for industrial engineering, the majority of postdoctoral appointees were supported by federal sources. In industrial engineering departments, 17 of the 27 postdoctoral appointees (63 percent) were supported by nonfederal sources.

Other nonfaculty research staff with doctorates. Engineering and computer science departments reported 731 nonfaculty research staff with doctorates in 1991, all but one of whom were at doctorate-granting institutions. Women represented 10 percent of these nonfaculty doctoral research staff. In general, there was less than one such staff member per engineering department at doctorate-granting institutions, the exception being departments of materials engineering. (See Table 5, Dickens Appendix, for the 1991 distribution of nonfaculty doctoral research staff.)

Student Enrollments in Engineering

Undergraduate students. One indicator of student awareness of career opportunities is changes in the preferences for majors and careers shown by first-year college students. According to data from an annual survey of incoming college students conducted since 1966, interest in engineering as a career has fluctuated, falling from 8.9 percent in 1966 to a low of 4.7 percent, then rising to a peak of 12.0 percent in 1982, followed by another decline to 8.1 percent in 1990. Women's interest in engineering careers rose from 0.2 percent in 1966 to a peak of 3.6 percent in 1982 then declined to 2.4 percent by 1990.[13] The proportion of underrepresented minority students—African Americans, Native Americans, and Hispanics—intending to major in engineering increased strongly over the past 20 years, rising from 7.3 percent in 1972 to 17.7 percent in 1992.[14] (See Table 6, Dickens Appendix, for data on career preferences of first-year college students.)

According to the Engineering Workforce Commission, full-time undergraduate engineering enrollment in the fall of 1992 was 344,126, an increase of 1.4 percent over the fall of 1991. (See Table 7, Dickens Appendix, for data on undergraduate engineering enrollment.) The enrollment of part-time undergraduates decreased by more than 5.4 percent to 38,399.

Total undergraduate engineering enrollment in 1992 was 382,525, an increase of more than 2,500 over 1991. Although the fall 1992 undergraduate enrollment in engineering was substantially below the fall 1983 level of 441,451, the mix of students was different. In 1983 there were 406,144 full-time students and 35,061 part-time students. For full-time students, the 1983 figure was the largest on record. For part-time students, however, the peak enrollment figure was the 41,445 recorded in the fall of 1990.[15]

The enrollment of women and underrepresented minorities continued to increase. In the fall of 1992, women represented over 19 percent of first-year students and over 17 percent of all full-time undergraduates. Underrepresented minorities (African Americans, Hispanic Americans, and Native Americans) increased their representation among first-year students to 17 percent and among all full-time undergraduates to over 13 percent. The representation of women and these minorities in 1992 were historically high levels.[16]

Graduate students. There are three sources of information on graduate engineering enrollments—the American Society for Engineering Education, the Engineering Workforce Commission, and the National Science Foundation. The NSF data are used for this section because they also provide information on graduate enrollment in computer science departments. It should be noted, however, that the NSF data include all computer science departments, not just those within engineering colleges.

In the fall of 1991, the NSF survey of graduate departments reported 149,135 graduate students in engineering and computer science, a record high level. (See Table 8, Dickens Appendix, for information on total graduate enrollment.) Between 1972 and 1991, total graduate enrollment in engineering departments increased by 171 percent. Departments in all fields experienced growth in graduate enrollment over this period, ranging from 65 percent for chemical engineering to 685 percent for computer science.

In the 1970s, the growth of part-time graduate enrollment was 89 percent, compared with 31 percent for full-time enrollment. All fields except aerospace engineering experienced growth. The increases ranged from 25 percent for materials engineering to 164 percent for computer science. Enrollment in aerospace engineering decreased by 26 percent overall, with declines in both full-time and part-time graduate students. (NSF did not collect data on graduate enrollment by gender during most of the 1970s.)

The graduate enrollment picture was different during the period 1980 to 1991. All fields experienced growth in enrollment, with increases of full-time students accounting for the larger part of the gain. Part-time graduate enrollment decreased, however, in chemical engineering departments over this period.

Between 1980 and 1991, overall engineering graduate enrollment increased by a much greater percentage for women than for men, but in 1991 men still accounted for a substantial majority (84 percent) of engineering graduate students in all fields.

The computer science departments had a somewhat different pattern of graduate enrollment growth over the 1980-1991 period. The overall increase in enrollment was 156 percent, with the growth in part-time enrollment exceeding that for full-time students. There was relatively little difference in the increase in full-time enrollments for women and men, but the increase in part-time enrollment for women was substantially greater than that for men. Over the 1980-1991 period, the proportion of graduate computer science enrollment represented by women increased slightly from 23 percent to 24 percent. (See Table 9, Dickens Appendix, for information on full-time and part-time graduate students.)

Degrees Awarded in Engineering and Computer Science

Bachelor's degrees. From 1966, when computer science degree data were first reported by the National Center for Education Statistics (NCES), the numbers of baccalaureates awarded in engineering and computer science increased each year until 1986, growing from 35,904 to 119,015 (an increase of 231 percent). Much of this growth resulted from the rapid rise in degrees in computer science and from the strong increases in the numbers of engineering degrees awarded to women. After 1986, however, the numbers of baccalaureates awarded each year in both engineering and computer science declined, with computer science having the sharper decrease.

When viewed by gender and field, the patterns were somewhat different. For women, engineering bachelor's degrees grew very rapidly until 1985, leveled off, and then began to decline. (Table 10, Dickens Appendix, presents data on bachelor's degrees by field and gender.) The peak year for engineering baccalaureates awarded to women was 1987 at 11,404, which was more than 78 times greater that the 146 degrees women earned in 1966. By 1990 baccalaureates awarded to women had decreased to 9,973, a decline of 13 percent from 1987. For men, there was a smaller overall rise, followed by a larger decline in the numbers of engineering bachelor's degrees. The growth in degrees awarded to men ended in 1985, two years earlier than for women. The 66,326 engineering baccalaureates awarded to men in 1985 was 86 percent above the figure for 1966. By 1990 the number of these degrees awarded to men had declined from the 1985 peak to 54,732, a decrease of 17 percent.

For computer science, the numbers of bachelor's degrees awarded to both men and women increased rapidly from 1966. For men, baccalaureates in computer science rose from 76 in 1966 to a peak of 27,069 in 1986 and then declined to 19,321 in 1990, a drop of 29 percent. The growth in women's baccalaureates in computer science was also very large, rising from 13 in 1966 to a peak of 15,126 in 1986. By 1990 the number of computer science bachelor's degrees awarded to women had dropped to 8,374, a decline of 45 percent from the 1986 figure. (Table 11, Dickens Appendix, presents data on master's degrees awarded by field and gender.)

The 219 engineering research universities awarded almost two-thirds (65 percent) of the engineering baccalaureates in 1990. There was considerable difference in this proportion by field of engineering, ranging from 49 percent for aeronautical engineering to 74 for materials engineering. In contrast, the engineering research universities awarded only about one-third (32.0 percent) of the bachelor's degrees in computer science. (See Table 1.)

Master's degrees. The number of master's degrees awarded in engineering and computer science fields grew dramatically over the 1966-1990 period, increasing from 13,916 to 33,638, a gain of 142 percent. The contribution of women to this increase is seen in their share of master's degrees, which rose from 0.7 percent in 1966 to 18 percent in 1990. Women earned 93 master's degrees in these fields in 1966 and 5,944 in 1990. For both men and women, the number of master's degrees awarded in these fields in 1990 was the largest over this period.

During the 1970s, the number of master's degrees awarded in many engineering fields declined. There followed a period of growth in the 1980s. Civil and chemical engineering reached their maximum numbers in 1984 and 1985, respectively. From their mid-decade peaks, the number of master's degrees in these fields declined by 10 percent for civil engineering and 34 percent for chemical engineering. Mechanical and materials engineering had their largest number of master's degrees in 1989 and had small decreases in 1990. In contrast, the number of master's degrees awarded in computer science increased throughout the 1966-1990 period.

Doctor's degrees. The 1966-1991 period may be divided into three distinct phases in terms of the number of doctorates awarded in engineering and computer science: (1) From 1966 to 1972 there was a large increase in these degrees, rising from 2,301 to 3,509; (2) between 1972 and 1978 a decline to 2,546 in the number of these doctorates erased most of the Phase One increase; and (3) from 1978 to 1991 there was a new period of growth, slow at first and then rapid after 1985. The total of 6,009 doctorates awarded in 1991 represents a new high record. The decline during the 1970s was accounted for by the drop in doctorates awarded to U.S. citizen and permanent resident males. Although their numbers continued to decline until 1982, the effect was offset by the strong growth in the number of doctorates awarded to foreign citizen males who were temporary residents of the United States. After 1982 doctorates awarded to U.S. citizen and permanent resident males began to increase again, helping fuel the growth in engineering and computer science degrees at this level. All fields of engineering shared in the growth. (See Table 12, Dickens Appendix, for data on doctorates awarded in engineering and computer science by citizenship and gender.)

Much of the growth in the number of engineering doctorates was accounted for by foreign students with temporary resident status; and, in 1991, for the first time, the number of doctorates awarded to temporary residents exceeded the number awarded to U.S. citizens and permanent residents. By field, the greatest increases in the number of doctorates awarded to temporary residents were in electrical engineering and mechanical engineering. In computer science, as well, foreign citizen temporary residents received an increasing share of doctor's degrees. In 1991 they received 42 percent of computer science doctorates, up from 20 percent in 1980.

The increasing number of foreign citizens among recipients of engineering and computer science doctorates from U.S. universities is also reflected in the nationality of the baccalaureate-origin institutions. In a special analysis, the National Science Foundation compared U.S. with foreign baccalaureate-origin institutions for doctorate recipients during the period from 1985 to 1990. The data are presented in Table 2.

Women were major contributors to the growth in doctorates in engineering and computer science between 1980 and 1991. The number of women receiving doctorates in engineering increased 402 percent from 1980 to 1991, with temporary residents gaining 813 percent, compared with 309 percent for U.S. citizens and permanent residents. As a result of this growth, the share of all engineering doctorates awarded to women increased from 4 percent in 1980 to 9 percent in 1991. By field, the largest number of doctorates awarded in 1991 to female temporary residents were in electrical engineering (30 or 22 percent); the leading fields for U.S. citizen and permanent resident females were other engineering (72 or 24 percent) and chemical engineering (60 or 20 percent). The number of computer science doctorates awarded to women grew by 452 percent between 1980 and 1991, compared with 242 percent for men. Women's share of computer science doctorates rose from 10 percent in 1980 to 15 percent in 1991.

FUNDING OF ACADEMIC ENGINEERING RESEARCH

Government Sources of Support for Academic Research by Field and Category of Research

All fields. In 1991, U.S. academic institutions reported overall R&D expenditures of approximately $17.2 billion for all fields, including engineering and computer science. (See Table 1 for a listing of these 219 institutions in rank order by R&D expenditures for 1991.) The activity distribution for these expenditures was basic research, 65.5 percent, applied research, 25.9 percent, and development, 8.6 percent. The sources of these funds are shown in Table 3.

R&D expenditures at engineering research universities. Table 4 summarizes data on sources of funding for the 219 universities and colleges that reported research and development expenditures for engineering and computer science in 1991. Of the total $3.1 billion in R&D expenditures, $2.64 billion (85 percent) was reported as engineering and $460 million (15 percent) was reported as computer science. These R&D expenditures were concentrated in a relatively small number of institutions, with 22 universities accounting for one-half of the total. Five universities reported R&D expenditures in engineering and computer science exceeding $100 million each; the Massachusetts Institute of Technology reported the largest such expenditure, $146 million.

The federal government provided about $1.7 billion, or 55 percent, of the R&D expenditures of engineering research universities in 1991. (See Table 5.) Federal sources provided 63 percent of the R&D expenditures for computer science, compared with 53 percent for engineering. Among the fields of engineering, the proportion of R&D expenditures that came from federal sources ranged from 39 percent for civil engineering to 72 percent for aerospace engineering.

Trends in R&D expenditures. From 1973 to 1991, R&D expenditures at engineering research universities increased, in constant 1989 dollars, by 264 percent for engineering and computer science. (See Table 13, Dickens Appendix, for data on R&D expenditures of academic institutions by field and source of funds.) Although the federal government provided the larger share of these funds over this period, their growth was much less than that for nonfederal sources, 198 percent and 430 percent, respectively.

Data by field of engineering for academic R&D expenditures, which became available beginning in 1980, reveal that there were increases in constant dollar terms for all fields of engineering from 1980 to 1991, ranging from 101 percent for mechanical engineering to 166 percent for chemical engineering. The increase for computer science was 197 percent.

Although federally funded R&D expenditures by engineering research universities grew between 1980 and 1991, in constant dollars, the increases were less than for total expenditures in all engineering fields and computer science. For engineering and computer science overall, the increase for federally funded R&D expenditures was 99 percent. For fields of engineering, the growth in federally funded R&D expenditures ranged from 57 percent for civil engineering to 114 percent for electrical engineering. The increase for computer science was 184 percent.

From 1980 to 1991, R&D expenditures funded by nonfederal sources at engineering research universities grew by an overall 218 percent, in constant dollars, for engineering and computer science. The increases for engineering fields ranged from 152 percent for mechanical engineering to 336 percent for civil engineering. The increase for computer science was 229 percent.

Support for graduate students as research assistants. Graduate research assistants provide a substantial part of the human resources that support R&D activities at universities in engineering and computer science. (See Table 14, Dickens Appendix, for data on sources of support for full-time graduate assistants by field.) Nonfederal sources have become increasingly important as the source of support for full-time graduate students who hold research assistantships in these fields. In 1972 the federal government supported almost two-thirds (62 percent) of the graduate research assistants in engineering and computer science. (See Table 6.) By 1991 less than one-half (46 percent) of the graduate research assistants were supported by federal sources. Nonfederal funding of graduate research assistants comes from many sources including the own funds of universities. See Table 4 for data on mechanisms of support for postdoctoral appointees, by field.

Support for postdoctoral appointees. In 1991 federal sources provided the support for 67 percent of the postdoctoral appointees in engineering departments and for 75 percent of those in computer science departments. The principal mechanism of federal support was through research grants, 94 percent for postdoctorates in engineering departments and 97 percent in computer science departments. In all engineering departments except industrial engineering, the majority of postdoctoral appointees were supported by federal sources. In industrial engineering departments, 17 of the 27 postdoctoral appointees (63 percent) were supported by nonfederal sources.

Engineering research centers and laboratories. In the NSF-funded study of organized engineering research units, Robert P. Morgan and coworkers made the following finding:

> The research units varied widely in size and research funding. About half of the units had annual engineering research expenditures of less than $1,000,000 while 5% had expenditures of $10 million or more. Individual units also differed widely in the sources of the support they received. Across all units, during FY 1992 the funding breakdown by source was as follows: U.S. federal government, 44.9%; U.S. business an industry, 22.6%; U.S. state and local government, 13.6%; internal university funds, 12.0%; foreign business, industry or government, 3.1%; other, (including private non-profit organizations, gifts, sales, etc.) 3.8%. Some 40% of those responding indicated that they received no internal university budget support during FY 1992.[17]

Mission Basis for Government Research Support

Government agencies support research and development activities as part of the fulfillment of their missions. A substantial part of the research and development work by universities is linked to the fulfillment of government agency missions. The annual budget proposal of the President requests budget authority from the Congress. The congressional appropriations determine what the budget authority will be. Support for research and development at universities in the fields of engineering and computer science comes from many budget function categories and federal agencies.

The fiscal year 1995 budget submitted by the President to the Congress in early February gives a clear indication of the Clinton/Gore administration's priorities for federal research and development investments. The most recently published figures on federal R&D budget authority are presented in Table 7. These data reflect the priorities of the Clinton/Gore administration in fiscal years 1994 and 1995 and those of the Bush administration in fiscal year 1993.

According to President Clinton's fiscal year 1995 budget,

> The administration is proposing $71 billion in R&D investments (excluding facilities) in 1995, a $2.5 billion, or 4 percent, increase over 1994. Civilian R&D will increase by more than $1 billion, or 4 percent, to $32 billion. The combination of continued annual growth for civilian R&D, anticipated decreases in defense R&D after 1995, and the inclusion of dual-use defense R&D is likely to cause the civilian share of the R&D budget to exceed 50 percent earlier than the 1998 date predicted in the 1994 budget. Much of this increase will be focused on cost-shared and competitively selected projects that are industry-defined and industry-led (i.e., consortia, cooperative R&D, etc.). In 1995 university-based research will increase to $12 billion, a $437 million, or 4 percent, increase over 1994. University-based research continues to provide an important contribution to the creation of knowledge, technological innovation, and the training of scientists and engineers.[18]

Industry Support for Academic Engineering Research

According to NSF data, in 1991 industry provided $1.2 billion, or about 7 percent, of the $17.6 billion spent by academic institutions on research and development activities. These data were not disaggregated by field, but other data indicated the proportion of industrial funding for engineering R&D was substantially greater than the overall average for all fields. Morgan and coworkers found that organized engineering research units at academic institutions received almost 23 percent of their funds from U.S. business and industry.[19] A National Research Council special study of chemical engineering noted that between 1980 and 1986, industrial support of academic research in that field nearly quadrupled and were the main force for funding growth in academic chemical engineering.[20]

Foreign Support of Academic Engineering Research

According to the Organization for Economic Cooperation and Development (OECD), foreign sources financed about 11 percent of the industrial R&D performance in the United States in 1991. Available data suggest that foreign sources provided a smaller share of R&D performance at U.S. academic institutions. The data collected by Morgan and others showed that foreign business, industry, or governments provided about 3 percent of the funding of organized engineering research units in U.S. universities.[21]

University Support of Academic Engineering Research with Own Funds

In 1991, according to NSF estimates, universities and colleges provided $4.9 billion for overall R&D activities and had expenditures of $17.6 billion for these purposes. Funds provided by universities and colleges represented 28 percent of their R&D expenditures. The NSF data included state and local government funds to the university and college sector. The study by Morgan and others found that in fiscal year 1992 the organized engineering research units received 12 percent of their support from internal university funds, and an additional 14 percent from state and local governments.[22] The sum of these two sources of funding in the Morgan study—about 26 percent—is roughly comparable to the figure reported in the NSF data.

NATURE AND SCOPE OF RELATIONSHIPS BETWEEN ENGINEERING RESEARCH AND EDUCATION

Undergraduate Students Participating in Research Programs

Early exposure to research is widely recognized as an important element in the development of future researchers. Many federal agencies provide support through special training programs, such as the National Science Foundation's Research Experiences for Undergraduates (REU), or through supplements to research grants for the addition of undergraduate students to the research team.

Organized research units at universities provide research experiences and employment to undergraduate engineering students. Morgan and coworkers found that 18 percent of the units reported a great extent of involvement of undergraduate students, and another 48 percent reported some involvement of these students. "Undergraduates most frequently were used as assistants to others in research followed by general 'go-fors' and by 'technicians'."[23] On average, each unit had 35 undergraduate students working, but a small number of centers with large expenditures tended to skew the data. If the 31 centers with annual research expenditures of more than $10 million were excluded, the average number of undergraduate students per center dropped from 35 to 22.[24]

Graduate Students Participating in Research Activities

In 1991, 35 percent of the full-time graduate students in engineering fields and computer science held research assistantships. (See Table 8.) The comparable proportion in 1972 was 19 percent. There were wide differences by field in the proportion of full-time graduate students who were supported by research assistantships. In 1972 only three fields supported more than one-fourth of their full-time graduate students as research assistants: materials engineering (44 percent), electrical engineering (27 percent), and chemical engineering (27 percent). In 1991 two fields had one-half or more of their full-time graduate students holding research assistantships (62 percent for materials engineering and 50 percent for chemical engineering).

Organized engineering research units at universities offer opportunities for research to graduate students. According to the study by Morgan and coworkers, the directors of 87 percent of the research units reported that there was a great extent of graduate student involvement, and another 10 percent of the directors reported some involvement. On average, each unit had 36 graduate students, but this figure is skewed by the small number of centers with large expenditures. If these 31 large centers with annual research expenditures of more than $10 million were excluded, the average number of graduate students per center dropped from 36 to 27.[25] "The most frequent roles for graduate students in research units were as 'associate researchers' followed next by the role of 'independent researcher.' . . . Approximately 87% of the graduate students working in research units are working on unit projects or problems that constitute their master's theses or doctoral dissertations."[26]

The Morgan study made the following observation about the contributions to graduate engineering education from student work in research units:

> Directors in our survey said that the most important ways that work in their research units adds to the education of graduate students and development of their engineering skills were as follows (in order of decreasing frequency): leads to easier entry into industry, provides cross-disciplinary research experience, leads to easier entry into academia, provides more focus on the problems of industry, and provides a better understanding of engineering's role in industry.[27]

Faculty, Postdoctorates, and Other Academic Engineers Participating in Research

University faculty are expected to maintain active participation in research as one of their primary duties. During the 1980s the structure of research support has changed, in part because of the increasing emphasis given by federal sponsoring agencies to interdisciplinary research. The traditional model for university research in many fields has been that of the individual investigator working with a small group of graduate students and postdoctoral students. A special study by the Federal Coordinating Council for Science, Engineering, and Technology found that the share of research funds going to individual investigators declined over the decade of the 1980s from 56 percent to 51 percent, while the shares for research teams and major facilities increased. Funding for research centers decreased slightly because of a slower growth rate in centers at the National Institutes of Health and the U.S. Department of Agriculture.[28]

The Morgan study reports similar findings. Organized engineering research units, on average, involved 28 faculty, research associates, postdoctoral students, and technical support personnel. The study concluded that

> There appears to be a shift taking place in university-based engineering research away from the individual investigator model towards more applied, team research of a cross-disciplinary nature. . . . Although changes in the nature and dimensions of university-based engineering research have occurred, the traditionally valued outputs of this research still predominate. In particular, when asked about the importance of a variety of research outputs, the research unit Directors specified that papers for publication, conference reports and presentations, and technical reports were of much greater importance than pieces of hardware, commercial or military products, and patents or invention disclosures. Thus, the more traditional academic outputs continue to predominate, even in an organized research setting in which more practical, applied research is being conducted.[29]

DATA CONSIDERATIONS

This section presents technical items related to the data used in the paper and raises points for consideration in planning future studies of this kind.

Definitions. "Engineering research universities" are those institutions that reported research and development expenditures for engineering or computer science in the 1991 National Science Foundation Survey of R&D Expenditures of Universities and Colleges. There are 219 such institutions.

For the purposes of this paper, the term *engineering* includes computer science. In most databases, computer science is reported separately from engineering.

Limitations. A long-term description of engineering research universities is hampered by the fact that prior to approximately 1980, the major federal sources for data on R&D funding did not provide disaggregated data for engineering fields. In contrast, data on degrees and enrollment by field of engineering are available for a longer period of time. Recently some of the data collected by the American Society for Engineering Education has been developed into a database.[30]

At the time the statistical tabulations were created for this paper, the most recent year for which federal data were generally available was 1991. Therefore, that year has been used for identification and characterization of engineering research universities.

Annual data are available for a number of major topics covered in the paper, including R&D expenditures of engineering research universities, obligations of federal agencies for research and development, engineering degrees awarded, and engineering enrollments. In contrast, data on engineering faculty are available only for certain years and only as aggregated tabulations.

Principal data sources. Much of the data presented in this paper was tabulated from the Computer Aided Science Policy Analysis and Research (CASPAR) Database System, developed by Quantum Research Corporation for the National Science Foundation (NSF). The most recently available version of CASPAR was released in June 1993 and includes data through 1991. CASPAR includes data from surveys of the NSF and the U.S. Department of Education (ED), as well as the National Research Council's Doctorate Records File. In addition, the report includes information from other NSF sources, including published reports and unpublished tabulations, the Engineering Workforce Commission, the Higher Education Research Institute of the University of California at Los Angeles, and a special study by Robert P. Morgan and others of the Washington University in St. Louis.

Data considerations for future studies. There is a need for the principal federal and nonfederal data collection organizations to increase the coverage and availability of data on engineering. Further questions that should be addressed include, but are not limited to, the following:

1. *Identifying a core set of data that is generally needed for policy studies in engineering.*

Most policy studies in a given sector, for example academia, use certain data sets to set the context. For the academic sector, there is usually concern about enrollments, degrees awarded, income and expenditures, faculty and other engineers, numbers of institutions, and numbers of departments in each engineering field. Of these familiar categories, there is a relative dearth of information on engineering faculty and other engineers employed in academia.

The core data set should also include information about the research activities of engineers in academia and about the organizational structures in which they perform this research. The current study being conducted by Robert Morgan and his associates may help illuminate these topics. What plans should be made to update some of this information on a regular basis?

In terms of resources for data collection, the core data set should not be allocated all the funds. There should be support for special studies that will address engineering issues not covered by the core data.

2. *Coordinating and setting priorities for data collection activities related to engineering.*

Data collection organizations would probably find it helpful to be able to discuss their current and planned activities with an identified, continuing body that could represent the interests and concerns of the engineering community. In the absence of such a group, there is the risk that decisions about engineering data collection will be overly influenced by the needs of some current topic or the views of just a few individuals, who may not represent the full scope of engineering concerns.

There is a need to form such a coordinating body to address the data collection activities of nonfederal as well as federal organizations. Because data collection is expensive and money is often tight, there is a need to set priorities in data collection activities. Moreover, the staffing reductions of industrial firms pose an added constraint on data collection from that sector. The views and recommendations of a coordinating body would be of great value in making decisions about what are the highest-priority data collection activities.

3. *Considering the unique problems of data for engineering policy studies.*

Engineering is a transcendent activity. Increasingly, its research activities are of an interdisciplinary or a multidisciplinary character, and the boundaries between academia, industry, and government are less and less distinct.

Any discussion of data needs for policy studies in engineering should consider the coordination of information collection activities across sectors. The coordinating body could help sector-specific data collection organizations develop ways to make their activities more useful for policy analyses that cut across sectors.

NOTES

1. It should be noted that the terminology in this paper is different from that of some widely used classifications in which, for example, the term "research universities" designates a subset of all universities that do research.

2. Robert P. Morgan, Donald E. Strickland, Nirmala Kannankutty, and Carol Spelman, "Engineering Research in U.S. Universities: How University-Based Research Directors See It," IEEE-ASEE Frontiers in Education Conference, November 7, 1993, p. 1.

3. *Ibid.*, p. 1.

4. *Ibid.*, p. 1.

5. National Research Council Transportation Research Board, "Measuring Quality: A Review Process for the University Transportation Centers Program," National Academy Press, Washington, DC, 1993, Table A-1.

6. National Science Foundation, "Engineering Research Centers: A Partnership for Competitiveness," NSF 92-48i, Arlington, VA 22230.

7. Richard W. Heckel, Professor of Metallurgical and Materials Engineering, College of Engineering, Michigan Technological University, Houghton, Michigan, has developed a database from the annual compilation by the American Society for Engineering Education (ASEE) of statistical information contributed by individual engineering programs. The ASEE data include degrees, enrollment, faculty, and research expenditures. The results of Professor Heckel's analyses are presented in two papers which have been submitted to *Engineering Education*. The titles of the draft papers are "Current and Emerging Statistical Trends in Engineering Education" and "Degrees and Research Funding in Various Engineering Disciplines Over the Last Two Decades."

8. A study on engineering faculty with disabilities conducted for the Society of Women in Engineering provided a total for engineering faculty in academic year 1992-93 of 21,374. According to a personal communication from Betty Vetter, Executive Director of the Commission on Professionals in Science and Technology, on December 15, 1993, the study did not provide data on total faculty by field of engineering.

9. U.S. Department of Education, National Survey of Postsecondary Faculty, 1987-88, as cited in National Center for Education Statistics *Digest of Education Statistics, 1991*, NCES 91-697, 221.

10. Frank J. Atelsek and Irene L. Gomberg, "Recruitment and Retention of Full-time Engineering Faculty, Fall 1980," *Higher Education Panel Reports*, American Council on Education, Washington, DC, October, 1981, p. 3.

11. National Science Foundation, "Survey of Research Participation and Characteristics of Science and Engineering Faculty, 1985-1986," Arlington, VA 22230, unpublished tabulations.

12. *Ibid.*

13. E. L. Dey, A. W. Astin, and W. S. Korn, *The American Freshman: Twenty-Five Year Trends, 1966-1990*, Higher Education Research Institute, University of California at Los Angeles, 1991.

14. Higher Education Research Institute, University of California at Los Angeles, Survey of the American Freshman: National Norms, Los Angeles, 1992, unpublished tabulations.

15. Engineering Workforce Commission, "Engineering Workforce Bulletin," American Association of Engineering Societies, Washington, DC, April 1993 and Engineering Manpower Commission, "Engineering Manpower Bulletin," American Association of Engineering Societies, April 1991, and April 1992.

16. Engineering Workforce Commission, "Engineering Workforce Bulletin," American Association of Engineering Societies, Washington, DC, April 1993.

17. Morgan, *op.cit.*, p. 1.

18. Office of Management and Budget, *Budget of the United States Government, Fiscal Year 1995*, "Chapter 3B. Investing for Productivity and Prosperity—Investing in Know How," Washington, DC 20506.

19. Morgan, *op. cit.*, p.1.

20. National Research Council, *Frontiers in Chemical Engineering: Research Needs and Opportunities*, National Academy Press, Washington, DC, 1988, p. 185.

21. Morgan, *op. cit.*, p. 1.

22. *Ibid.*

23. *Ibid.*, p. 4.

24. Robert P. Morgan, comments on the draft version of the Background Paper, February 14, 1994.

25. Robert P. Morgan, *ibid.*

26. *Ibid.*, p. 4.

27. *Ibid.*, p. 4.

28. Federal Coordinating Council for Science, Engineering, and Technology, *Trends in the Structure of Federal Science Support*, Office of Science and Technology Policy, Washington, DC, 1992, p. 2-7.

29. Morgan, *op.cit.*, p. 5.

30. An alternative source of data has been developed from the annual statistical information provided by individual engineering programs to the American Society for Engineering Education by Professor Richard W. Heckel, Department of Metallurgical and Materials Engineering, Michigan Technological University for the period Academic Year 1970 to Academic Year 1990. Professor Heckel has submitted his analyses of these data for publication by *Engineering Education.*

APPENDIX

TABLE 1 Total Research and Development Expenditures for Engineering and Computer Science by Academic Institution, 1991 (in thousands of dollars)

	Academic institution	Total	Engineering	Computer science
	All academic institutions	3,437,214	2,892,750	544,464
1	Massachusetts Institute of Technology*	146,038	132,421	13,617
2	Georgia Institute of Technology, All Campuses	141,785	124,708	17,077
3	Pennsylvania State U, All Campuses	121,744	120,336	1,408
4	Stanford University*	101,064	92,089	8,975
5	University of Texas at Austin	100,981	86,521	14,460
6	University of Michigan, All Campuses	74,294	63,841	10,453
7	University of Illinois at Urbana-Champaign	72,761	54,694	18,067
8	Texas A&M University, All Campuses	70,674	65,627	5,047
9	Carnegie Mellon University*	69,705	24,615	45,090
10	Cornell University*, All Campuses	64,923	43,632	21,291
	First 10 institutions	963,969	808,484	155,485
11	University of California-Berkeley	59,132	57,481	1,651
12	University of Southern California*	56,826	24,010	32,816
13	University of Minnesota, All Campuses	52,646	27,460	25,186
14	New Mexico State University, All Campuses	52,448	44,091	8,357
15	Virginia Polytechnic Institute and State Univ	49,143	48,407	736
16	North Carolina State University at Raleigh	48,201	46,343	1,858
17	University of Wisconsin-Madison	47,795	43,341	4,454
18	Iowa State University	46,013	36,768	9,245
19	Utah State University	45,156	44,842	314
20	Ohio State University, All Campuses	44,305	42,950	1,355
	First 20 institutions	1,465,634	1,224,177	241,457
21	University of Maryland at College Park	43,437	36,868	6,569
22	Rensselaer Polytechnic Institute*	39,520	37,927	1,593
23	Purdue University, All Campuses	38,483	35,130	3,353
24	University of Tennessee Central Office	36,947	30,817	6,130
25	University of Dayton*	34,108	33,660	448
26	Louisiana State University, All Campuses	32,048	31,768	280
27	University of Florida	31,832	29,445	2,387
28	University of California-Los Angeles	29,323	25,715	3,608
29	Rutgers, the State University, All Campuses	28,889	24,591	4,298
30	SUNY at Buffalo, All Campuses	27,080	22,881	4,199
	First 30 institutions	1,807,301	1,532,979	274,322
31	University of Colorado, All Campuses	25,577	21,550	4,027
32	Princeton University*	23,893	20,134	3,759
33	University of Rochester*	23,825	22,304	1,521
34	Case Western Reserve University*	23,770	23,770	0
35	Clemson University	23,406	18,715	4,691

36	University of Connecticut, All Campuses	23,280	21,575	1,705
37	University of California-Santa Barbara	23,206	20,648	2,558
38	Northwestern University*	22,977	19,603	3,374
39	Auburn University, All Campuses	22,855	22,182	673
40	University of Washington	22,747	20,497	2,250
	First 40 institutions	2,042,837	1,743,957	298,880
41	California Institute of Technology*	22,724	17,931	4,793
42	University of New Mexico, All Campuses	22,477	22,083	394
43	Arizona State University	21,262	20,129	1,133
44	University of Arizona	20,999	18,080	2,919
45	Columbia University*, Main Campus	20,721	16,785	3,936
46	Lehigh University*	20,576	20,185	391
47	University of Virginia, All Campuses	20,517	17,748	2,769
48	University of Utah	18,616	15,116	3,500
49	University of Pennsylvania*	17,524	10,620	6,904
50	University of Massachusetts, All Campuses	17,305	9,717	7,588
	First 50 institutions	2,245,558	1,912,351	333,207
51	University of South Florida	16,945	16,945	0
52	West Virginia University	16,834	7,005	9,829
53	Colorado State University	16,819	16,406	413
54	University of Cincinnati, All Campuses	16,129	15,961	168
55	University of Oklahoma, All Campuses	15,964	13,724	2,240
56	University of Alabama in Huntsville	15,570	12,781	2,789
57	Mississippi State University	15,460	15,360	100
58	Michigan State University	15,340	15,340	0
59	University of South Carolina, All Campuses	15,062	14,260	802
60	University of Kentucky, All Campuses	14,863	14,610	253
	First 60 institutions	2,404,544	2,054,743	349,801
61	Syracuse University*, All Campuses	14,634	5,255	9,379
62	University of Houston-University Park	14,544	14,396	148
63	University of Delaware	13,976	12,528	1,448
64	University of California-San Diego	13,794	5,624	8,170
65	Oklahoma State University, All Campuses	13,737	13,203	534
66	Brown University*	13,622	9,936	3,686
67	University of Nebraska at Lincoln	13,554	12,377	1,177
68	Rice University*	13,057	4,304	8,753
69	Drexel University*	12,675	12,506	169
70	Michigan Technological University	12,455	12,326	129
	First 70 institutions	2,540,592	2,157,198	383,394
71	Woods Hole Oceanographic Institution*	12,165	12,165	0
72	University of Iowa	12,027	10,805	1,222
73	Johns Hopkins University*	11,311	10,800	511
74	New Mexico Institute of Mining and Technology	11,302	11,239	63
75	Yale University*	10,731	7,329	3,402
76	University of Missouri, Rolla	10,626	10,286	340
77	University of California-Davis	10,269	9,634	635
78	University of Missouri, Columbia	9,952	9,937	15
79	Texas Tech University	9,429	8,902	527
80	Oregon State University	9,163	8,226	937
	First 80 institutions	2,647,567	2,256,521	391,046

TABLE 1 Continued

	Academic institution	Total	Engineering	Computer science
81	Colorado School of Mines	9,128	8,648	480
82	University of Texas at Arlington	9,090	8,240	850
83	Vanderbilt University*	9,008	8,554	454
84	University of California-Irvine	8,991	5,171	3,820
85	University of North Dakota, All Campuses	8,361	8,361	0
86	Duke University*	8,272	6,512	1,760
87	Tennessee Technological University	8,267	8,242	25
88	University of Georgia	8,220	4,312	3,908
89	University of Idaho	8,082	8,031	51
90	University of Illinois at Chicago	7,940	7,940	0
	First 90 institutions	2,732,926	2,330,532	402,394
91	Washington University*	7,786	2,953	4,833
92	University of Pittsburgh, All Campuses	7,569	6,514	1,055
93	University of North Carolina at Chapel Hill	7,569	0	7,569
94	Stevens Institute of Technology*	7,469	7,469	0
95	SUNY College of Environmental Science and Forestry	7,432	7,139	293
96	Southern Illinois University-Carbondale	7,348	5,134	2,214
97	University of Alabama	7,327	5,996	1,331
98	Northeastern University*	7,243	5,925	1,318
99	Mercer University*, All Campuses	7,184	7,184	0
100	New Jersey Institute Technology	7,093	6,614	479
	First 100 institutions	2,806,946	2,385,460	421,486
101	New York University*	7,029	0	7,029
102	Washington State University	6,809	6,792	17
103	Kansas State University of Agriculture and App Sci	6,760	6,635	125
104	Polytechnic University*	6,683	6,683	0
105	Ohio University, All Campuses	6,678	6,678	0
106	University of Notre Dame*	6,583	6,583	0
107	Wayne State University	6,475	5,624	851
108	Clarkson University*	6,370	6,370	0
109	Howard University*	6,240	3,871	2,369
110	North Carolina Agricultural and Technical St Univ	6,230	5,753	477
	First 110 institutions	2,872,803	2,440,449	432,354
111	University of Central Florida	6,142	5,418	724
112	University of Arkansas, Main Campus	5,986	5,710	276
113	San Diego State University	5,915	4,466	1,449
114	Brigham Young University*, All Campuses	5,531	5,358	173
115	George Mason University	5,355	5,109	246
116	Illinois Institute of Technology*	5,309	5,093	216
117	University of Akron, All Campuses	5,197	5,197	0
118	Institute of Paper Science and Technology*	5,161	4,595	566

119	George Washington University*	5,123	5,123	0
120	Boston University*	5,096	4,677	419
	First 120 institutions	2,927,618	2,491,195	436,423
121	Dartmouth College*	5,078	5,078	0
122	University of Rhode Island	4,930	4,708	222
123	CUNY City College	4,921	4,834	87
124	University of Alabama at Birmingham	4,874	4,711	163
125	Oregon Graduate Inst. of Science & Technology*	4,836	2,948	1,888
126	Tulane University of Louisiana*	4,749	4,317	432
127	Worcester Polytechnic Institute	4,686	4,397	289
128	University of Lowell	4,602	3,925	677
129	Old Dominion University	4,513	3,758	755
130	Cleveland State University	4,452	4,410	42
	First 130 institutions	2,975,259	2,534,281	440,978
131	University of Kansas, All Campuses	4,401	4,365	36
132	University of Texas at El Paso	4,349	3,922	427
133	Montana State University	4,340	4,330	10
134	Wichita State University	4,252	4,149	103
135	SUNY at Stony Brook, All Campuses	4,210	2,640	1,570
136	Florida Atlantic University	3,850	3,074	776
137	University of Tulsa*	3,767	3,570	197
138	University of Maine at Orono	3,763	3,724	39
139	Harvard University*	3,701	2,346	1,355
140	University of Alaska Fairbanks, All Campuses	3,622	3,616	6
	First 140 institutions	3,015,514	2,570,017	445,497
141	University of Nevada-Reno	3,465	3,465	0
142	University of New Hampshire, Main Campus	3,415	3,098	317
143	Wright State University, All Campuses	3,383	2,408	975
144	San Jose State University	3,114	3,114	0
145	University of Wisconsin-Milwaukee	3,078	2,998	80
146	Florida Agricultural and Mechanical University	2,832	2,797	35
147	University of Mississippi, All Campuses	2,762	2,762	0
148	University of Wyoming	2,650	2,603	47
149	Tufts University*	2,321	2,307	14
150	University of Toledo	2,282	2,230	52
	First 150 institutions	3,044,816	2,597,799	447,017
151	Lamar University-Beaumont	2,252	2,148	104
152	North Dakota State University, All Campuses	2,230	1,902	328
153	California Polytechnic State Univ-San Luis Obispo	2,083	1,844	239
154	Louisiana Tech University	2,015	2,015	0
155	University of North Carolina at Charlotte	1,985	1,799	186
156	Indiana University, All Campuses	1,915	656	1,259
157	University of North Texas	1,906	502	1,404
158	Florida Institute of Technology*	1,849	1,696	153
159	Jackson State University	1,826	1,345	481
160	University of Miami*	1,816	1,643	173
	First 160 institutions	3,064,693	2,613,349	451,344

TABLE 1 Continued

	Academic institution	Total	Engineering	Computer science
161	University of Hawaii at Manoa	1,805	1,607	198
162	Western Michigan University	1,779	1,631	148
163	University of Puerto Rico Mayaguez	1,682	1,673	9
164	Memphis State University	1,603	1,397	206
165	University of Maryland Baltimore County	1,547	1,384	163
166	University of Oregon	1,457	939	518
167	Catholic University of America*	1,443	1,443	0
168	Marquette University*	1,375	1,375	0
169	University of Missouri, Kansas City	1,289	0	1,289
170	Southern University and A & M Col, All Campuses	1,272	423	849
	First 170 institutions	3,079,945	2,625,221	454,724
171	Southern Methodist University*	1,243	922	321
172	SUNY at Binghamton	1,234	1,014	220
173	University of Louisville	1,234	1,207	27
174	Florida State University	1,210	846	364
175	University of California-Santa Cruz	1,085	373	712
176	University of Texas at Dallas	1,064	738	326
177	University of Vermont	997	997	0
178	University of Chicago*	985	0	985
179	Georgetown University*	980	0	980
180	Portland State University	957	804	153
	First 180 institutions	3,090,934	2,632,122	458,812
181	South Dakota School of Mines & Technology	899	899	0
182	Santa Clara University*	829	644	185
183	Oakland University	736	673	63
184	South Dakota State University	674	674	0
185	Kent State University, All Campuses	670	150	520
186	SUNY at Albany	650	0	650
187	Tuskegee University*	644	644	0
188	College of William and Mary, All Campuses	631	0	631
189	Northern Illinois University	625	600	25
190	Tennessee State University	483	483	0
	First 190 institutions	3,097,775	2,636,889	460,886
191	Milwaukee School of Engineering*	432	432	0
192	Temple University	429	282	147
193	Ball State University	398	187	211
194	Central State University	386	347	39
195	Hampton University*	334	259	75
196	Northern Arizona University	331	331	0
197	Brandeis University*	331	0	331
198	Georgia State University	287	0	287
199	Morgan State University	244	244	0
200	University of Denver*	240	240	0
	First 200 institutions	3,101,187	2,639,211	461,976

201	Hofstra University*	214	214	0
202	Indiana State University, All Campuses	175	150	25
203	West Virginia State College	170	170	0
204	Prairie View A&M University	157	157	0
205	Virginia Commonwealth University	129	120	9
206	Eastern Washington University	128	0	128
207	University of South Alabama	126	120	6
208	University of the District of Columbia	104	104	0
209	Miami University, All Campuses	79	79	0
210	Alabama Agricultural and Mechanical University	67	67	0
	First 210 institutions	3,102,536	2,640,392	462,144
211	American University*	61	0	61
212	Canisius College*	43	0	43
213	University of California-Riverside	40	40	0
214	Boston College*	34	0	34
215	Texas Southern University	34	0	34
216	Northeast Louisiana University	15	0	15
217	Stephen F Austin State University	12	0	12
218	CUNY Queens College	6	0	6
219	University of South Dakota	5	0	5
	First 219 institutions	3,102,786	2,640,432	462,354

*Privately controlled institutions.
SOURCE: NSF CASPAR Database System.

TABLE 2 Science and Engineering Doctorate-Holders Employed by Academic Institutions and Those Active in Research and Development (R&D), 1979 and 1989

Field	Total employment		Total in R&D			
			Number		Percent	
	1979	1989	1979	1989	1979	1989
Engineering and computer science	16,031	27,607	12,150	21,871	75.8	79.2
Engineering, total	13,839	21,517	10,659	17,749	77.0	82.5
Aerospace	598	1,031	556	893	93.0	86.6
Chemical	1,060	2,051	777	1,886	73.3	92.0
Civil	2,165	3,278	1,822	2,529	84.2	77.2
Electrical	2,490	4,402	1,830	3,442	73.5	78.2
Materials	1,300	1,595	1,044	1,421	80.3	89.1
Mechanical	2,374	3,988	1,675	3,295	70.6	83.4
Other	3,852	5,222	2,955	4,283	76.7	82.2
Computer science	2,192	6,090	1,491	4,122	68.0	67.7

SOURCE: National Science Foundation, *Science and Engineering Indicators - 1991*, Appendix table 5-20, Washington, DC, p. 375.

TABLE 3 Postdoctorates in Graduate Engineering and Computer Science Departments by Field, Citizenship, and Gender, 1980-1991

Field	1980	1981	1982	1983	1984	1985	1986	1987	1988	1989	1990	1991
Total postdoctorates												
Engineering and computer science	1,021	1,072	1,025	1,187	1,261	1,425	1,476	1,547	1,784	1,998	2,018	2,406
Engineering, total	978	1,037	978	1,105	1,197	1,349	1,399	1,443	1,688	1,914	1,947	2,249
Aerospace	20	14	25	32	42	51	48	43	48	38	67	77
Chemical	189	173	179	199	249	279	296	319	431	475	563	596
Civil	122	103	103	131	146	122	140	175	203	182	168	186
Electrical	123	191	178	178	173	177	173	176	187	193	242	307
Mechanical	137	130	130	182	196	207	240	216	218	304	220	331
Materials	172	194	168	204	168	245	250	283	325	323	370	401
Industrial	16	13	9	13	21	18	25	26	32	32	6	27
Other	199	219	186	166	202	250	227	205	244	367	311	324
Computer science	43	35	47	82	64	76	77	104	96	84	71	157
Total U.S. citizen postdoctorates												
Engineering and computer science	332	349	358	450	480	488	507	566	644	699	643	722
Engineering, total	302	331	323	413	437	441	459	497	587	652	603	612
Aerospace	6	8	6	8	10	10	15	13	20	14	21	29
Chemical	47	43	44	56	66	56	63	68	100	119	161	132
Civil	59	33	37	52	53	49	46	57	91	80	77	69
Electrical	39	104	82	70	63	77	60	104	71	58	78	71
Mechanical	43	24	30	69	82	61	94	69	77	101	66	102
Materials	27	26	30	53	52	47	56	71	72	86	89	80
Industrial	8	6	4	6	12	16	19	19	23	17	3	9
Other	73	87	90	99	99	125	106	96	133	177	108	120
Computer science	30	18	35	37	43	47	48	69	57	47	40	110

Total foreign citizen postdoctorates												
Engineering and computer science	689	723	667	737	781	937	969	981	1,140	1,299	1,375	1,684
Engineering, total	676	706	655	692	760	908	940	946	1,101	1,262	1,344	1,637
Aerospace	14	6	19	24	32	41	33	30	28	24	46	48
Chemical	142	130	135	143	183	223	233	251	331	356	402	464
Civil	63	70	66	79	93	73	94	118	112	102	91	117
Electrical	84	87	96	108	110	100	113	72	116	135	164	236
Mechanical	94	106	100	113	114	146	146	147	141	203	154	229
Materials	145	168	138	151	116	198	194	212	253	237	281	321
Industrial	8	7	5	7	9	2	6	7	9	15	3	18
Other	126	132	96	67	103	125	121	109	111	190	203	204
Computer science	13	17	12	45	21	29	29	35	39	37	31	47
Total female postdoctorates												
Engineering and computer science	67	88	93	98	93	111	142	162	184	192	218	269
Engineering, total	65	82	84	88	83	101	131	148	172	177	209	242
Aerospace	0	2	0	0	0	1	2	4	3	3	4	4
Chemical	21	11	22	27	31	28	38	35	53	46	82	94
Civil	10	8	6	6	10	3	7	12	19	29	14	23
Electrical	5	10	17	12	11	11	13	16	16	13	15	18
Mechanical	3	10	5	6	7	14	21	11	19	15	16	22
Materials	12	16	16	24	14	23	23	45	36	36	44	43
Industrial	0	3	1	2	1	3	4	1	4	6	1	2
Other	14	22	17	11	9	18	23	24	22	29	33	36
Computer science	2	6	9	10	10	10	11	14	12	15	9	27

TABLE 3 Continued

Field	1980	1981	1982	1983	1984	1985	1986	1987	1988	1989	1990	1991
Female U.S. citizen postdoctorates												
Engineering and computer science	34	42	55	54	53	50	72	84	75	80	82	87
Engineering, total	32	38	47	51	46	43	67	71	65	67	75	74
Aerospace	0	0	0	0	0	0	2	4	2	1	1	2
Chemical	11	5	11	10	14	9	17	11	13	12	20	21
Civil	5	2	5	6	7	3	4	4	9	12	7	12
Electrical	2	5	8	6	6	6	7	14	10	5	9	4
Mechanical	0	4	4	5	5	6	13	5	5	8	6	9
Materials	5	7	8	15	8	7	8	17	10	11	18	11
Industrial	0	3	1	1	1	2	3	1	4	5	1	2
Other	9	12	10	8	5	10	13	15	12	13	13	13
Computer science	2	4	8	3	7	7	5	13	10	13	7	13
Female foreign citizen postdoctorates												
Engineering and computer science	33	46	38	44	40	61	70	78	109	112	136	182
Engineering, total	33	44	37	37	37	58	64	77	107	110	134	168
Aerospace	0	2	0	0	0	1	0	0	1	2	3	2
Chemical	10	6	11	17	17	19	21	24	40	34	62	73
Civil	5	6	1	0	3	0	3	8	10	17	7	11
Electrical	3	5	9	6	5	5	6	2	6	8	6	14
Mechanical	3	6	1	1	2	8	8	6	14	7	10	13
Materials	7	9	8	9	6	16	15	28	26	25	26	32
Industrial	0	0	0	1	0	1	1	0	0	1	0	0
Other	5	10	7	3	4	8	10	9	10	16	20	23
Computer science	0	2	1	7	3	3	6	1	2	2	2	14

SOURCE: NSF CASPAR Database System.

TABLE 4 Postdoctoral Appointees in Doctorate-Granting Institutions by Field and Source of Support: 1991

		Source of support				
	Total	Federal				Nonfederal
Field		Total	Fellow-ships	Trainee-ships	Research grants	
Engineering and computer science	2,394	1,605	69	25	1,511	789
Engineering, total	2,237	1,488	65	25	1,398	749
Aerospace	77	55	0	0	55	22
Agriculture*	33	18	1	0	17	15
Biomedical*	66	53	10	8	35	13
Chemical	578	323	7	3	313	255
Civil	185	131	6	0	125	54
Electrical	300	202	6	3	193	98
Engineering science*	117	93	9	0	84	24
Industrial	27	10	1	0	9	17
Mechanical	329	237	23	7	207	92
Materials	401	273	2	2	269	128
Mining*	11	8	0	0	8	3
Nuclear*	29	22	0	0	22	7
Petroleum*	18	16	0	0	16	2
Engineering, n.e.c.*	66	47	0	2	45	19
Computer science	157	117	4	0	113	40

KEY: *indicates fields included in "other engineering" in other tables.
n.e.c. = not elsewhere classified

SOURCE: National Science Foundation, Division of Science Resources Studies.

TABLE 5 Nonfaculty Doctoral Research Staff at Doctorate-Granting Institutions by Field and Gender, 1991

Field	Number of departments	Non-faculty doctoral research staff	
		Total	Women
Engineering and computer science	1,494	731	71
Engineering, subtotal	1,260	682	59
Aerospace	44	26	1
Chemical	157	74	15
Civil	192	54	3
Electrical	199	120	7
Industrial	134	20	8
Mechanical	177	139	5
Materials	95	146	10
Other	262	103	10
Computer science	234	49	12

SOURCE: National Science Foundation, Division of Science Resources Studies.

TABLE 6 Percent of U.S. College Freshmen Choosing Engineering as a Career, by Sex: 1966-1990

Year	All	Male	Female
1966	8.9	16.3	.2
1967	8.4	15.0	.2
1968	8.3	14.6	.2
1969	8.3	14.5	.3
1970	7.5	13.3	.4
1971	5.3	9.7	.2
1972	5.3	9.6	.3
1973	5.3	9.4	.7
1974	4.7	8.5	.8
1975	5.9	10.2	1.1
1976	7.8	13.7	1.5
1977	8.3	15.1	1.5
1978	9.1	16.5	2.2
1979	9.3	16.8	2.3
1980	10.7	19.1	2.9
1981	10.9	19.5	2.9
1982	12.0	20.6	3.6
1983	10.8	18.8	3.3
1984	10.4	18.5	2.9
1985	10.0	17.7	2.9
1986	9.7	17.4	2.8
1987	8.5	15.2	2.6
1988	8.6	15.7	2.5
1989	9.0	16.5	2.6
1990	8.1	14.9	2.4

SOURCE: E. L. Day, A. W. Astin, and W. S. Korn, The American Freshman: Twenty-Five Year Trends, 1966-1990, Los Angeles: Higher Education Research Institute, UCLA, 1991.

TABLE 7 Fall Engineering Enrollments of Undergraduates by Status and Class Year, 1980-1992

	1980	1981	1982	1983	1984	1985	1986	1987	1988	1989	1990	1991	1992
Total	397,344	420,402	435,330	441,205	429,499	420,864	407,657	392,198	385,412	378,277	380,287	379,977	382,525
Full-time, subtotal	365,117	387,577	403,390	406,144	394,635	384,191	369,520	356,998	346,169	338,529	338,842	339,397	344,126
First year	110,149	115,280	115,303	109,638	105,249	103,225	99,238	95,453	98,009	95,420	94,346	93,002	93,427
Second year	84,982	87,519	89,785	89,515	83,946	79,627	76,195	73,317	71,030	71,267	72,204	71,257	71,644
Third year	80,024	86,633	90,541	91,233	89,509	84,875	80,386	77,085	73,761	70,483	72,666	73,516	74,871
Fourth year	84,442	92,414	102,055	109,036	109,695	110,305	107,773	104,003	97,614	94,465	92,989	94,683	98,235
Fifth year	5,520	5,731	5,706	6,722	6,236	6,159	5,928	7,140	5,755	6,894	6,637	6,939	5,949
Part-time, subtotal	32,227	32,825	31,940	35,061	34,864	36,673	38,137	35,200	39,243	39,748	41,445	40,580	38,399
Gender													
Male	345,482	361,133	368,750	372,374	362,800	354,612	344,999	331,917	325,024	318,067	318,471	316,441	316,460
Female	51,862	59,269	66,580	68,831	66,699	66,252	62,658	60,281	60,388	60,210	61,816	63,536	66,065
Race/ethnicity													
White	326,913	343,649	356,750	354,329	340,374	323,899	315,861	296,749	288,415	281,948	288,732	271,906	270,942
Asian	12,772	15,815	17,570	23,007	25,449	28,767	30,201	32,795	34,051	33,360	30,898	37,803	38,480
Underrep. minorities	31,531	34,353	35,960	37,432	37,557	39,657	37,240	38,640	40,389	41,338	41,169	48,692	51,517
Black	17,606	18,911	19,400	19,698	19,204	19,819	18,459	19,142	20,405	21,013	20,833	24,563	25,722
Hispanic	12,905	14,359	15,320	16,462	17,075	18,598	17,586	18,253	18,700	19,007	18,873	22,441	23,863
American Indian	1,020	1,083	1,240	1,272	1,278	1,240	1,195	1,245	1,284	1,318	1,463	1,688	1,932
Temporary resident	26,128	26,585	25,050	26,437	26,119	28,541	24,355	24,014	22,557	21,631	19,488	21,576	21,586

SOURCES: American Society of Engineering Societies, *Manpower Comments*, April 1991 and April 1992; *Engineering Workforce Bulletin*, April 1993.

TABLE 8 Graduate Enrollment in Engineering and Computer Science Departments by Field, Enrollment Status, and Gender, 1972-1991

Field	1972	1973	1974	1975	1976	1977	1978	1979
All students								
Engineering and computer science	55,027	56,385	63,422	76,775	75,377	77,890	72,144	83,497
Engineering, subtotal	50,602	51,233	56,977	68,360	66,750	68,782	63,815	71,807
Aerospace	2,014	1,816	1,654	1,670	1,477	1,518	1,450	1,481
Chemical	4,740	4,718	5,011	5,336	5,581	5,580	5,611	6,029
Civil	7,954	8,673	10,115	12,560	11,995	12,352	11,565	12,836
Electrical	13,325	13,713	15,530	16,320	15,926	17,406	16,379	17,715
Mechanical	6,309	7,293	8,107	8,601	8,313	8,722	8,122	9,251
Materials	2,211	2,030	2,156	2,352	2,375	2,559	2,487	2,756
Industrial	5,507	5,208	6,141	11,663	10,687	10,438	8,967	10,714
Other	8,542	7,782	8,263	9,858	10,396	10,207	9,234	11,025
Computer science	4,425	5,152	6,445	8,415	8,627	9,108	8,329	11,690
All full-time students								
Engineering and computer science	35,092	34,164	37,483	42,286	41,698	41,829	41,147	45,870
Engineering, subtotal	32,191	31,226	33,737	37,813	36,950	37,225	36,721	40,017
Aerospace	1,538	1,362	1,245	1,245	1,165	1,187	1,135	1,152
Chemical	3,435	3,442	3,569	3,743	4,014	4,174	4,226	4,555
Civil	5,267	5,367	5,939	7,363	7,025	7,111	7,123	7,637
Electrical	7,612	7,462	7,769	8,278	8,147	8,528	8,334	9,039
Mechanical	3,978	4,405	4,712	4,931	4,919	4,883	4,868	5,428
Materials	1,720	1,619	1,638	1,787	1,860	1,951	1,961	2,135
Industrial	3,047	2.518	3,108	4,152	3,578	3,343	3,140	3,743
Other	5,594	5,051	5,757	6,314	6,242	6,048	5,934	6,328
Computer science	2,901	2,938	3,746	4,473	4,748	4,604	4,426	5,853

TABLE 8 Continued

Field	1972	1973	1974	1975	1976	1977	1978	1979
All part-time students								
Engineering and computer science	19,935	22,221	25,939	34,489	33,679	36,061	30,997	37,627
Engineering, subtotal	18,411	20,007	23,240	30,547	29,800	31,557	27,094	31,790
Aerospace	476	454	409	425	312	331	315	329
Chemical	1,305	1,276	1,442	1,593	1,567	1,406	1,385	1,474
Civil	2,687	3,306	4,176	5,197	4,970	5,241	4,442	5,199
Electrical	5,713	6,251	7,761	8,042	7,779	8,878	8,045	8,676
Mechanical	2,331	2,888	3,395	3,670	3,394	3,839	3,254	3,823
Materials	491	411	518	565	515	608	526	621
Industrial	2,460	2,690	3,033	7,511	7,109	7,095	5,827	6,971
Other	2,948	2,731	2,506	3,544	4,154	4,159	3,300	4,697
Computer science	1,524	2,214	2,699	3,942	3,879	4,504	3,903	5,837
All female students								
Engineering and computer science	1,221	1,337	1,864	2,461	2,768	5,264	0	8,208
Engineering, subtotal	851	918	1,359	1,870	2,026	3,705	0	5,885
Aerospace	16	14	15	20	23	33	0	49
Chemical	95	98	119	166	243	388	0	645
Civil	131	160	303	434	513	830	0	1,323
Electrical	142	187	221	228	297	710	0	933
Mechanical	39	54	81	112	122	273	0	469
Materials	46	52	70	99	109	183	0	288
Industrial	131	117	237	478	379	755	0	1,251
Other	251	236	313	333	340	533	0	927
Computer science	370	419	505	591	742	1,559	0	2,323

Female full-time students								
Engineering and computer science	1,221	1,337	1,864	2,461	2,768	3,020	0	4,488
Engineering, subtotal	851	918	1,359	1,870	2,026	2,261	0	3,438
Aerospace	16	14	15	20	23	27	0	36
Chemical	95	98	119	166	243	289	0	458
Civil	131	160	303	434	513	543	0	898
Electrical	142	187	221	228	297	382	0	454
Mechanical	39	54	81	112	122	141	0	258
Materials	46	52	70	99	109	142	0	213
Industrial	131	117	237	478	379	376	0	579
Other	251	236	313	333	340	361	0	542
Computer science	370	419	505	591	742	759	0	1,050

	1980	1981	1982	1983	1984	1985	1986	1987	1988	1989	1990	1991
All students												
Engineering and computer science	88,043	96,195	103,710	114,727	118,590	126,004	133,537	136,026	135,713	136,917	141,702	149,135
Engineering, subtotal	74,465	79,758	83,898	91,111	92,780	96,160	102,140	103,889	103,159	104,319	107,329	114,407
Aerospace	1,737	1,883	1,941	2,305	2,340	2,538	2,804	3,015	3,223	3,454	3,866	4,041
Chemical	6,518	7,017	7,775	8,300	8,117	7,932	7,759	7,844	7,360	7,125	7,327	7,838
Civil	13,111	14,103	14,146	14,921	15,203	14,916	14,987	14,718	14,822	14,919	15,454	17,265
Electrical	19,132	20,113	21,927	25,116	26,198	28,026	29,799	31,214	31,837	33,055	33,583	35,272
Mechanical	9,888	10,618	11,467	12,911	13,855	14,157	15,713	16,278	16,207	16,239	16,455	17,820
Materials	2,910	3,125	3,124	3,447	3,657	3,943	4,208	4,366	4,335	4,589	4,921	5,149
Industrial	9,737	9,797	9,641	9,373	9,535	10,841	11,888	12,457	11,731	11,458	11,611	13,333
Other	11,432	13,102	13,877	14,738	13,875	13,807	14,982	13,997	13,644	13,480	14,112	13,689
Computer science	13,578	16,437	19,812	23,616	25,810	29,844	31,397	32,137	32,554	32,598	34,373	34,728

TABLE 8 Continued

	1980	1981	1982	1983	1984	1985	1986	1987	1988	1989	1990	1991
All full-time students												
Engineering and computer science	49,307	53,296	59,036	64,618	66,744	70,039	75,537	77,457	78,580	80,343	82,551	87,782
Engineering, subtotal	42,720	45,851	49,865	53,931	55,157	55,938	60,227	61,885	63,187	64,546	65,692	71,230
Aerospace	1,327	1,483	1,522	1,840	1,854	1,994	2,154	2,372	2,533	2,772	3,010	3,314
Chemical	4,863	5,382	6,084	6,489	6,348	6,123	6,166	6,264	5,958	5,788	5,853	6,309
Civil	7,964	8,805	9,398	9,807	10,089	9,760	9,982	9,651	9,957	9,974	10,074	11,214
Electrical	9,963	10,450	11,533	13,203	13,857	14,752	16,222	17,040	17,653	18,396	18,567	19,988
Mechanical	5,841	6,422	7,267	8,330	8,698	8,875	9,789	10,164	10,426	10,464	10,662	11,654
Materials	2,254	2,454	2,478	2,704	2,872	3,093	3,370	3,436	3,464	3,715	3,922	4,059
Industrial	3,764	3,639	3,827	3,322	3,433	3,517	3,875	4,233	4,398	4,761	4,863	5,705
Other	6,744	7,216	7,756	8,236	8,006	7,824	8,669	8,725	8,798	8,676	8,741	8,987
Computer science	6,587	7,445	9,171	10,687	11,587	14,101	15,310	15,572	15,393	15,797	16,859	16,552
All part-time students												
Engineering and computer science	38,736	42,899	44,674	50,109	51,846	55,965	58,000	58,569	57,133	56,574	59,151	61,353
Engineering, subtotal	31,745	33,907	34,033	37,180	37,623	40,222	41,913	42,004	39,972	39,773	41,637	43,177
Aerospace	410	400	419	465	486	544	650	643	690	682	856	727
Chemical	1,655	1,635	1,691	1,811	1,769	1,809	1,593	1,580	1,402	1,337	1,474	1,529
Civil	5,147	5,298	4,748	5,114	5,114	5,156	5,005	5,067	4,865	4,945	5,380	6,051
Electrical	9,169	9,663	10,394	11,913	12,341	13,274	13,577	14,174	14,184	14,659	15,016	15,284
Mechanical	4,047	4,196	4,200	4,581	5,157	5,282	5,924	6,114	5,781	5,775	5,793	6,166
Materials	656	671	646	743	785	850	838	930	871	874	999	1,090
Industrial	5,973	6,158	5,814	6,051	6,102	7,324	8,013	8,224	7,333	6,697	6,748	7,628
Other	4,688	5,886	6,121	6,502	5,869	5,983	6,313	5,272	4,846	4,804	5,371	4,702
Computer science	6,991	8,992	10,641	12,929	14,223	15,743	16,087	16,565	17,161	16,801	17,514	18,176

All female students												
Engineering and computer science	9,427	11,963	14,239	16,439	17,246	18,659	20,249	20,908	21,411	21,610	22,871	24,132
Engineering, subtotal	6,340	7,754	8,793	9,791	10,341	11,160	12,465	12,962	13,207	13,675	14,704	15,856
Aerospace	74	67	110	123	132	163	200	224	227	240	287	325
Chemical	720	830	943	1,088	1,032	1,070	1,087	1,194	1,124	1,057	1,175	1,312
Civil	1,366	1,576	1,752	1,860	2,028	2,149	2,151	2,113	2,297	2,438	2,684	3,164
Electrical	983	1,276	1,551	2,045	2,188	2,409	2,890	3,076	3,255	3,491	3,698	3,875
Mechanical	534	631	719	805	956	1,011	1,159	1,283	1,415	1,403	1,426	1,527
Materials	303	390	420	448	468	603	675	729	756	809	895	985
Industrial	1,350	1,544	1,631	1,544	1,691	1,903	2,224	2,300	2,190	2,188	2,216	2,402
Other	1,010	1,440	1,667	1,878	1,846	1,852	2,079	2,043	1,943	2,049	2,323	2,266
Computer science	3,087	4,209	5,446	6,648	6,905	7,499	7,784	7,946	8,204	7,935	8,167	8,276
Female full-time students												
Engineering and computer science	5,263	6,353	7,567	8,260	8,806	9,376	10,352	10,653	11,028	11,457	12,197	13,287
Engineering, subtotal	3,860	4,626	5,256	5,611	6,044	6,284	6,973	7,284	7,644	8,176	8,690	9,751
Aerospace	54	51	73	80	86	126	148	167	168	180	199	254
Chemical	492	567	705	832	784	801	826	913	823	816	916	1,026
Civil	900	1,127	1,243	1,295	1,438	1,494	1,491	1,404	1,568	1,652	1,780	2,041
Electrical	558	711	818	955	1,146	1,199	1,457	1,542	1,674	1,814	1,945	2,172
Mechanical	316	347	433	483	561	559	627	724	835	855	838	946
Materials	231	287	308	324	354	439	521	569	579	638	693	744
Industrial	667	695	771	619	630	647	748	753	756	926	929	1,059
Other	642	841	905	1,023	1,045	1,019	1,155	1,212	1,241	1,295	1,390	1,509
Computer science	1,403	1,727	2,311	2,649	2,762	3,092	3,379	3,369	3,384	3,281	3,507	3,536

TABLE 8 Continued

	1980	1981	1982	1983	1984	1985	1986	1987	1988	1989	1990	1991
Female part-time students												
Engineering and computer science	4,164	5,610	6,672	8,179	8,440	9,283	9,897	10,255	10,383	10,153	10,674	10,845
Engineering, subtotal	2,480	3,128	3,537	4,180	4,297	4,876	5,492	5,678	5,563	5,499	6,014	6,105
Aerospace	20	16	37	43	46	37	52	57	59	60	88	71
Chemical	228	263	238	256	248	269	261	281	301	241	259	286
Civil	466	449	509	565	590	655	660	709	729	786	904	1,123
Electrical	425	565	733	1,090	1,042	1,210	1,433	1,534	1,581	1,677	1,753	1,703
Mechanical	218	284	286	322	395	452	532	559	580	548	588	581
Materials	72	103	112	124	114	164	154	160	177	171	202	241
Industrial	683	849	860	925	1,061	1,256	1,476	1,547	1,434	1,262	1,287	1,343
Other	368	599	762	855	801	833	924	831	702	754	933	757
Computer science	1,684	2,482	3,135	3,999	4,143	4,407	4,405	4,577	4,820	4,654	4,660	4,740

SOURCE: NSF CASPAR Database System.

TABLE 9 Change in Graduate Enrollment in Engineering and Computer Science Departments by Full-Time and Part-Time Status and by Gender, 1980-1991

	Change in enrollment (percent)				
Field	Total	Full-time	Part-time	Women	Men
Engineering and computer science	69.4	78.0	58.4	156.0	59.0
Engineering, subtotal	53.6	66.7	36.0	150.0	44.7
Aeronautical	132.6	149.7	77.3	339.2	123.5
Chemical	20.3	29.7	-7.6	82.2	12.5
Civil	31.7	40.8	17.6	131.6	20.1
Electrical	84.4	100.6	66.7	73.0	294.2
Mechanical	80.2	99.5	52.4	186.0	74.2
Materials	76.9	80.1	66.2	225.1	59.7
Industrial	36.9	51.6	27.7	77.9	30.3
Other	19.7	33.2	0.3	224.4	9.6
Computer science	155.8	151.3	160.0	168.1	152.1

SOURCE: NSF CASPAR Database System.

TABLE 10 Bachelor's Degrees Awarded in Engineering and Computer Science by Field and Gender, 1966-1990

Field	1966	1967	1968	1969	1970	1971	1972	1973	1974	1975	1976	1977	1978	1979
Total bachelor's degrees														
Engineering and computer science	35,915	36,419	38,139	42,515	46,314	47,636	49,113	51,084	48,005	44,863	44,454	47,783	54,475	62,238
Engineering, total	35,826	36,197	37,680	41,582	44,770	45,248	45,711	46,779	43,248	39,824	38,790	41,357	47,251	53,469
Aerospace	1,683	1,914	2,072	2,625	2,756	2,443	2,180	1,738	1,210	1,174	1,009	1,078	1,186	1,386
Chemical	2,981	2,997	3,395	3,768	3,995	3,907	3,967	3,968	3,826	3,420	3,543	3,986	5,205	6,442
Civil	5,611	5,439	5,796	6,282	6,800	6,939	7,258	8,013	8,633	8,289	8,493	8,898	9,900	10,583
Electrical	11,007	10,843	10,725	11,695	12,288	12,288	12,181	12,377	11,419	10,246	9,874	10,018	11,213	12,440
Mechanical	7,811	7,890	7,930	8,514	9,310	9,177	8,784	8,795	7,883	7,089	6,984	7,927	9,100	10,360
Materials	792	836	881	952	977	916	909	885	821	711	704	738	835	1,045
Industrial	2,335	2,366	2,727	3,000	3,199	3,210	3,713	3,508	2,921	2,583	2,241	2,264	2,712	2,804
Other	3,606	3,912	4,154	4,746	5,445	6,368	6,719	7,495	6,535	6,312	5,942	6,448	7,100	8,409
Computer science	89	222	459	933	1,544	2,388	3,402	4,305	4,757	5,039	5,664	6,426	7,224	8,769

Field	1966	1967	1968	1969	1970	1971	1972	1973	1974	1975	1976	1977	1978	1979
Women														
Engineering and computer science	159	208	271	433	536	685	953	1,216	1,478	1,801	2,441	3,583	5,346	7,344
Engineering, total	146	184	216	312	337	361	492	576	698	845	1,317	2,044	3,482	4,881
Aerospace	5	14	12	19	20	17	20	18	18	24	29	28	61	66
Chemical	23	28	30	53	57	64	80	94	120	147	289	452	752	1,055
Civil	23	28	31	47	53	60	77	89	145	173	279	485	759	1,049
Electrical	29	42	43	66	68	76	82	158	117	130	193	268	435	659
Mechanical	19	20	32	40	39	43	49	63	66	84	150	242	472	620
Materials	7	8	18	10	10	13	16	15	32	35	43	59	107	183
Industrial	10	9	15	16	21	20	40	31	44	59	87	149	323	428
Other	30	35	35	61	69	68	128	108	156	193	247	361	573	821
Computer science	13	24	55	121	199	324	461	640	780	956	1,124	1,539	1,864	2,463

Field	1980	1981	1982	1983	1984	1985	1986	1987	1988	1989	1990
Total bachelor's degrees											
Engineering and computer science	70,023	78,950	87,891	97,352	108,588	116,693	119,015	114,352	105,050	97,910	92,400
Engineering, total	58,810	63,717	67,460	72,670	76,153	77,572	76,820	74,425	70,154	66,947	64,705
Aerospace	1,424	1,809	2,120	2,127	2,534	2,854	2,902	2,989	3,092	2,944	3,048
Chemical	7,276	7,639	8,059	8,550	9,192	8,941	7,411	6,114	4,654	4,187	3,834
Civil	11,046	11,331	11,280	10,747	10,351	9,730	9,223	8,746	8,131	8,015	7,992
Electrical	13,902	15,040	16,553	19,205	21,541	23,668	26,112	26,791	25,942	24,318	23,015
Mechanical	12,020	13,573	14,315	16,031	17,040	17,200	16,586	15,723	15,331	15,217	14,693
Materials	1,303	1,434	1,696	1,392	1,355	1,276	1,259	1,152	1,211	1,114	1,166
Industrial	3,217	3,878	4,044	3,824	4,020	4,009	4,255	4,313	4,259	4,121	4,041
Other	8,622	9,013	9,393	10,794	10,120	9,894	9,072	8,597	7,534	7,031	6,916
Computer science	11,213	15,233	20,431	24,682	32,435	39,121	42,195	39,927	34,896	30,963	27,695
Women											
Engineering and computer science	9,351	12,016	15,390	18,644	22,795	25,677	26,264	25,293	22,132	19,733	18,347
Engineering, total	5,952	7,063	8,275	9,652	10,729	11,246	11,138	11,404	10,779	10,188	9,973
Aerospace	82	129	171	172	175	241	248	248	298	301	343
Chemical	1,287	1,365	1,612	1,789	2,077	2,093	1,606	1,540	1,132	1,170	1,089
Civil	1,087	1,231	1,318	1,484	1,423	1,342	1,229	1,196	1,171	1,174	1,262
Electrical	902	1,100	1,411	1,922	2,289	2,732	3,227	3,564	3,524	3,188	2,867
Mechanical	893	1,151	1,266	1,485	1,812	1,801	1,710	1,727	1,764	1,680	1,715
Materials	227	270	324	288	322	286	335	298	320	261	271
Industrial	545	767	952	1,000	1,071	1,167	1,281	1,384	1,245	1,261	1,206
Other	929	1,050	1,221	1,512	1,560	1,584	1,502	1,447	1,325	1,153	1,220
Computer science	3,399	4,953	7,115	8,992	12,066	14,431	15,126	13,889	11,353	9,545	8,374

TABLE 10 Continued

Field	1980	1981	1982	1983	1984	1985	1986	1987	1988	1989	1990
Men											
Engineering and computer science	60,672	66,934	72,501	78,708	85,793	91,016	92,751	89,059	82,918	78,177	74,053
Engineering, total	52,858	56,654	59,185	63,018	65,424	66,326	65,682	63,021	59,375	56,759	54,732
Aerospace	1,342	1,680	1,949	1,955	2,359	2,613	2,654	2,741	2,794	2,643	2,705
Chemical	5,989	6,274	6,447	6,761	7,115	6,848	5,805	4,574	3,522	3,017	2,745
Civil	9,959	10,100	9,962	9,263	8,928	8,388	7,994	7,550	6,960	6,841	6,730
Electrical	13,000	13,940	15,142	17,283	19,252	20,936	22,885	23,227	22,418	21,130	20,148
Mechanical	11,127	12,422	13,049	14,546	15,228	15,399	14,876	13,996	13,567	13,537	12,978
Materials	1,076	1,164	1,372	1,104	1,033	990	924	854	891	853	895
Industrial	2,672	3,111	3,092	2,824	2,949	2,842	2,974	2,929	3,014	2,860	2,835
Other	7,693	7,963	8,172	9,282	8,560	8,310	7,570	7,150	6,209	5,878	5,696
Computer science	7,814	10,280	13,316	15,690	20,369	24,690	27,069	26,038	23,543	21,418	19,321

SOURCE: NSF CASPAR Database System.

TABLE 11 Master's Degrees Awarded in Engineering and Computer Science by Field and Gender, 1966-1990

Field	1966	1967	1968	1969	1970	1971	1972	1973	1974	1975	1976	1977	1978	1979
Total master's degrees														
Engineering and computer science	13,943	14,370	15,780	16,260	17,056	17,955	18,741	18,658	17,481	17,466	18,648	18,810	19,118	18,334
Engineering, total	13,705	13,921	15,232	15,248	15,597	16,367	16,764	16,545	15,205	15,167	16,045	16,012	16,080	15,279
Aerospace	798	802	841	835	749	717	687	563	557	477	479	385	411	372
Chemical	1,072	1,028	1,251	1,227	1,127	1,200	1,259	1,139	1,111	1,078	1,129	1,179	1,335	1,276
Civil	2,218	2,225	2,435	2,426	2,503	2,700	2,869	3,195	3,247	3,268	3,605	3,606	3,226	3,165
Electrical	3,872	3,953	4,226	4,033	4,138	4,282	4,209	3,899	3,499	3,471	3,774	3,788	3,742	3,596
Mechanical	2,154	2,176	2,136	2,299	2,298	2,502	2,552	2,396	2,058	2,032	2,088	2,094	2,095	2,012
Materials	400	444	460	441	429	480	524	582	521	500	475	504	506	529
Industrial	1,200	1,341	1,512	1,453	1,763	1,921	1,731	1,595	1,734	1,687	1,751	1,609	1,722	1,502
Other	1,991	1,952	2,371	2,534	2,590	2,565	2,933	3,176	2,478	2,654	2,744	2,847	3,043	2,827
Computer science	238	449	548	1,012	1,459	1,588	1,977	2,113	2,276	2,299	2,603	2,798	3,038	3,055
Women														
Engineering and computer science	93	104	129	186	304	350	496	503	640	710	945	1,164	1,410	1,512
Engineering, total	76	78	99	113	169	186	271	278	347	372	568	698	843	937
Aerospace	6	2	6	7	3	6	7	2	9	7	10	8	11	17
Chemical	7	7	14	11	18	27	29	22	31	27	41	69	90	120
Civil	9	21	17	20	30	44	48	60	83	107	151	185	196	214
Electrical	22	11	22	22	29	30	52	49	55	58	104	134	142	143
Mechanical	7	7	6	4	12	7	25	22	27	20	32	55	66	73
Materials	3	1	2	6	6	8	11	13	13	17	28	23	38	54
Industrial	6	10	9	8	17	23	25	29	45	56	81	75	138	128
Other	16	19	23	35	54	41	74	81	84	80	121	149	162	188
Computer science	17	26	30	73	135	164	225	225	293	338	377	466	567	575

TABLE 11 Continued

Field	1980	1981	1982	1983	1984	1985	1986	1987	1988	1989	1990
Total master's degrees											
Engineering and computer science	19,590	20,669	22,492	24,207	26,335	28,073	29,166	30,551	31,892	33,142	33,638
Engineering, total	15,943	16,451	17,557	18,886	20,145	20,972	21,096	22,070	22,726	23,743	23,995
Aerospace	382	408	521	491	562	605	621	737	797	855	1,029
Chemical	1,393	1,406	1,409	1,545	1,798	1,814	1,641	1,386	1,322	1,321	1,205
Civil	3,198	3,428	3,456	3,504	3,551	3,542	3,281	3,267	3,134	3,296	3,213
Electrical	3,842	3,902	4,465	4,819	5,519	5,649	6,147	6,895	7,455	7,849	8,009
Mechanical	2,194	2,419	2,539	2,683	2,964	3,272	3,256	3,380	3,513	3,703	3,630
Materials	598	666	632	672	726	713	810	765	749	815	802
Industrial	1,313	1,631	1,656	1,432	1,557	1,463	1,653	1,728	1,816	1,823	1,834
Other	3,023	2,591	2,879	3,740	3,468	3,914	3,687	3,912	3,940	4,081	4,273
Computer science	3,647	4,218	4,935	5,321	6,190	7,101	8,070	8,481	9,166	9,399	9,643
Women											
Engineering and computer science	1,887	2,300	2,885	3,263	3,911	4,281	4,812	5,266	5,272	5,708	5,944
Engineering, total	1,123	1,329	1,575	1,755	2,100	2,244	2,400	2,770	2,808	3,082	3,269
Aerospace	9	20	39	37	27	31	43	55	63	64	82
Chemical	144	176	187	176	208	285	240	243	215	229	192
Civil	265	316	352	382	415	414	373	475	413	445	520
Electrical	184	221	288	335	438	495	639	717	813	916	991
Mechanical	107	127	151	166	199	228	254	247	295	326	354
Materials	59	79	72	105	121	113	137	165	152	181	152
Industrial	133	166	210	206	278	227	279	319	324	358	341
Other	222	224	276	348	414	451	435	549	533	563	637
Computer science	764	971	1,310	1,508	1,811	2,037	2,412	2,496	2,464	2,626	2,675

SOURCE: NSF CASPAR Database System.

TABLE 12 Doctorates Awarded in Engineering and Computer Science by Field, Citizenship, and Gender, 1966-1991

Field	1966	1967	1968	1969	1970	1971	1972	1973	1974	1975	1976	1977	1978	1979
Total doctorates														
Engineering and computer science	2,301	2,607	2,864	3,276	3,446	3,514	3,509	3,375	3,161	3,011	2,868	2,679	2,546	2,704
Engineering, total	2,301	2,607	2,864	3,276	3,446	3,514	3,509	3,374	3,161	3,011	2,838	2,648	2,425	2,494
Aerospace	109	142	166	197	204	198	181	167	148	141	122	115	103	81
Chemical	367	330	377	422	457	407	391	424	418	396	335	329	282	315
Civil	293	307	368	364	366	427	437	435	390	361	388	336	303	302
Electrical	569	675	741	829	857	862	815	787	678	714	711	667	539	611
Mechanical	457	537	597	646	635	611	616	541	544	487	417	372	377	366
Materials	211	267	215	280	303	306	294	299	280	272	252	248	247	236
Industrial	46	61	74	111	117	134	142	109	92	92	67	73	51	82
Other	249	288	326	427	507	569	633	612	611	548	546	508	523	501
Computer science	0	0	0	0	0	0	0	1	0	0	0	31	121	210
U.S. citizens and permanent residents														
Engineering and computer science	1,834	2,157	2,384	2,742	2,952	2,960	2,958	2,707	2,271	2,137	1,949	1,824	1,676	1,792
Engineering, total	1,834	2,157	2,384	2,742	2,952	2,960	2,958	2,706	2,271	2,137	1,949	1,799	1,586	1,617
Aerospace	84	125	143	171	181	171	157	138	118	112	92	68	62	45
Chemical	308	284	311	357	400	340	332	326	282	268	212	215	179	193
Civil	195	224	268	255	270	315	343	312	242	231	229	210	170	178
Electrical	451	570	641	723	760	754	706	637	500	497	500	475	356	421
Mechanical	377	451	510	547	557	530	534	449	424	359	303	251	248	242
Materials	166	223	192	234	266	252	237	234	192	201	165	172	177	150
Industrial	41	44	60	102	88	121	124	96	61	72	40	48	40	54
Other	212	236	259	353	430	477	525	514	452	397	408	360	354	334
Computer science	0	0	0	0	0	0	0	1	0	0	0	25	90	175

TABLE 12 Continued

Field	1966	1967	1968	1969	1970	1971	1972	1973	1974	1975	1976	1977	1978	1979
Total doctorates to females														
Engineering and computer science	8	9	12	10	16	16	22	47	34	52	55	79	64	89
Engineering, total	8	9	12	10	16	16	22	46	34	52	55	74	53	62
Aerospace	0	0	1	0	2	3	0	2	3	2	0	3	1	0
Chemical	2	3	3	3		3	4	2	4	9	5	8	10	5
Civil	0	1	0	1	1	1	2	9	3	5	6	8	8	4
Electrical	2	0	0	3	3	4	5	7	3	16	15	21	17	11
Mechanical	1	2	2	0	2	0	2	7	7	4	4	6	3	5
Materials	2	1	2	1	1	1	3	7	3	5	8	10	5	8
Industrial	0	2	1	1	0	2	0	2	1	2	2	5	2	5
Other	1	0	3	1	4	1	8	8	5	13	12	11	12	20
Computer science	0	0	0	0	0	0	0	1	0	0	0	5	11	27
Female U.S. citizens and permanent residents														
Engineering and computer science	5	8	6	7	16	11	21	31	24	39	42	61	43	71
Engineering, total	5	8	6	7	16	11	21	30	24	39	42	57	36	44
Aerospace	0	0	1	0	2	2	0	2	3	1	0	1	1	0
Chemical	1	2	2	2	3	3	2	2	4	3	6	6	2	6
Civil	0	1	0	0	1	0	2	5	2	5	5	8	6	1
Electrical	0	0	0	3	3	3	4	3	1	10	12	16	12	8
Mechanical	1	2	0	0	2	0	2	5	6	2	3	5	1	5
Materials	2	1	1	1	1	1	3	4	2	4	6	8	3	5
Industrial	0	2	0	1	0	1	0	2	1	1	2	3	2	4
Other	1	0	2	0	4	1	8	7	5	13	8	10	9	15
Computer science	0	0	0	0	0	0	0	1	0	0	0	4	7	27

Field	1980	1981	1982	1983	1984	1985	1986	1987	1988	1989	1990	1991
Total doctorates												
Engineering and computer science	2,697	2,760	2,866	3,067	3,208	3,476	3,775	4,162	4,703	5,156	5,598	6,009
Engineering, total	2,479	2,528	2,646	2,781	2,913	3,166	3,376	3,712	4,188	4,544	4,893	5,212
Aerospace	81	97	86	106	119	124	118	142	150	178	192	207
Chemical	316	317	333	392	409	504	531	584	685	712	658	690
Civil	306	358	368	397	408	391	429	477	532	539	553	573
Electrical	540	549	616	625	660	716	806	779	1,010	1,137	1,276	1,405
Mechanical	384	360	437	379	427	513	536	657	715	760	883	874
Materials	273	234	255	268	271	303	305	392	374	380	440	490
Industrial	77	66	79	86	84	92	101	120	127	162	151	163
Other	502	547	472	528	535	523	550	561	595	676	740	810
Computer science	218	232	220	286	295	310	399	450	515	612	705	797
U.S. citizens and permanent residents												
Engineering and computer science	1,723	1,659	1,620	1,689	1,708	1,807	1,975	2,188	2,473	2,625	2,743	2,799
Engineering, total	1,554	1,471	1,465	1,482	1,513	1,594	1,726	1,913	2,147	2,229	2,340	2,358
Aerospace	44	56	44	53	57	70	45	72	81	82	90	107
Chemical	179	177	188	215	217	292	298	350	408	444	391	392
Civil	179	192	180	187	201	170	192	203	234	247	241	203
Electrical	359	350	335	349	356	353	410	393	503	539	597	657
Mechanical	243	202	263	192	202	251	267	308	327	325	367	338
Materials	175	135	131	145	150	165	152	214	199	194	215	235
Industrial	44	42	49	45	36	38	47	57	51	59	69	63
Other	331	317	275	296	294	255	315	316	344	339	370	363
Computer science	169	188	155	207	195	213	249	275	326	396	403	441

TABLE 12 Continued

Field	1980	1981	1982	1983	1984	1985	1986	1987	1988	1989	1990	1991
Total doctorates to females												
Engineering and computer science	111	125	144	160	188	231	273	307	342	483	525	568
Engineering, total	90	99	124	124	151	198	225	242	286	375	415	452
Aerospace	1	0	1	2	2	5	1	10	9	8	4	7
Chemical	14	11	19	23	27	41	61	60	65	80	78	80
Civil	11	10	17	13	25	20	21	18	30	54	49	39
Electrical	17	22	22	13	15	35	38	32	48	67	84	77
Mechanical	7	6	17	8	15	26	18	17	29	29	38	57
Materials	14	17	17	30	26	32	24	45	33	45	49	74
Industrial	7	6	6	6	16	6	14	13	19	18	25	17
Other	19	27	25	29	25	33	48	47	53	74	88	101
Computer science	21	26	20	36	37	33	48	65	56	108	110	116

Female U.S. citizens and permanent residents

Engineering and computer science	94	92	105	121	128	160	205	228	242	359	366	395
Engineering, total	74	68	88	91	101	136	162	174	196	275	276	303
Aerospace	1	0	0	2	1	5	1	6	6	5	1	4
Chemical	9	9	13	16	18	31	42	52	53	69	62	60
Civil	10	7	11	11	15	11	15	11	15	40	34	24
Electrical	15	16	14	12	9	22	25	15	33	36	47	42
Mechanical	5	4	15	6	12	18	13	11	17	19	24	36
Materials	10	11	9	20	16	22	18	36	19	36	35	53
Industrial	6	6	5	4	11	4	10	10	13	12	20	12
Other	18	15	21	20	19	23	38	33	40	58	53	72
Computer science	20	24	17	30	27	24	43	54	46	84	90	92

NOTE: The 1991 total includes records of 177 individuals for whom gender was not available.The distribution by field is as follows: engineering total, 170; aerospace, 10; chemical, 10; civil, 18; electrical, 37; mechanical, 35; materials, 7; industrial, 7; other engineering, 46; and computer science, 7.

SOURCE: NSF CASPAR Database System.

TABLE 13 Research and Development Expenditures of Academic Institutions by Field and Source of Funds, 1973-1991

Field	1973	1974	1975	1976	1977	1978	1979
All sources							
(Current dollars in thousands)							
Engineering and computer science	368,786	386,107	426,505	476,230	554,036	656,295	873,254
Engineering, subtotal	333,129	346,905	380,912	431,727	498,473	591,962	775,553
Computer science	35,657	39,202	45,593	44,503	55,563	64,333	97,701
(Constant 1989 dollars in thousands)							
Engineering and computer science	941,021	902,963	908,424	953,222	1,039,662	1,147,971	1,403,269
Engineering, subtotal	850,036	811,284	811,314	864,145	935,397	1,035,442	1,246,269
Computer science	90,985	91,679	97,110	89,077	104,265	112,529	157,000
Federal sources							
(Current dollars in thousands)							
Engineering and computer science	263,068	268,057	293,228	323,443	374,271	442,107	602,053
Engineering, subtotal	238,139	239,346	259,353	290,518	336,725	402,102	532,763
Computer science	24,929	28,711	33,875	32,925	37,546	40,005	69,290
(Constant 1989 dollars in thousands)							
Engineering and computer science	671,263	626,888	624,554	647,404	702,329	773,320	967,464
Engineering, subtotal	607,652	559,743	552,403	581,501	631,873	703,344	856,119
Computer science	63,611	67,145	72,151	65,903	70,456	69,976	111,345
Nonfederal sources							
(Current dollars in thousands)							
Engineering and computer science	105,718	118,050	133,277	152,787	179,765	214,188	271,201
Engineering, subtotal	94,990	107,559	121,559	141,209	161,748	189,860	242,790
Computer science	10,728	10,491	11,718	11,578	18,017	24,328	28,411
(Constant 1989 dollars in thousands)							
Engineering and computer science	269,758	276,075	283,870	305,818	337,333	374,651	435,805
Engineering, subtotal	242,384	251,541	258,911	282,644	303,524	332,098	390,150
Computer science	27,374	24,534	24,959	23,174	33,809	42,553	45,65

Field	1980	1981	1982	1983	1984	1985	1986
All sources							
(Current dollars in thousands)							
Engineering and computer science	986,587	1,110,884	1,191,718	1,308,643	1,456,082	1,698,684	1,962,463
Engineering, subtotal	862,351	966,996	1,028,024	1,122,372	1,231,950	1,417,876	1,641,081
Aerospace	53,096	54,397	62,386	68,473	69,757	80,525	94,422
Chemical	60,762	85,946	89,132	95,989	101,513	116,210	132,260
Civil	83,202	108,611	115,893	126,513	139,714	153,156	178,090
Electrical	183,146	193,292	218,459	261,960	295,159	337,403	394,984
Mechanical	140,378	140,773	142,743	149,388	178,975	207,751	228,117
Materials	0	0	0	0	0	0	0
Other	341,767	383,977	399,411	420,049	446,832	522,831	613,208
Computer science	124,236	143,888	163,694	186,271	224,132	280,808	321,382
(Constant 1989 dollars in thousands)							
Engineering and computer science	1,454,071	1,492,522	1,505,074	1,590,862	1,707,613	1,934,500	2,178,094
Engineering, subtotal	1,270,967	1,299,202	1,298,338	1,364,420	1,444,764	1,614,709	1,821,400
Aerospace	78,255	73,085	78,790	83,240	81,807	91,704	104,797
Chemical	89,553	115,472	112,569	116,690	119,049	132,343	146,792
Civil	122,626	145,924	146,367	153,796	163,849	174,417	197,658
Electrical	269,928	259,696	275,902	318,454	346,146	384,242	438,384
Mechanical	206,895	189,135	180,277	181,605	209,892	236,592	253,182
Materials	0	0	0	0	0	0	0
Other	503,710	515,890	504,434	510,636	524,020	595,412	680,586
Computer science	183,104	193,320	206,737	226,442	262,850	319,790	356,695
Nonfederal sources							
(Current dollars in thousands)							
Engineering and computer science	307,161	344,069	379,348	434,790	504,483	635,862	752,224
Engineering, subtotal	270,395	304,376	337,097	387,458	443,271	550,771	663,503
Aerospace	10,877	10,868	13,029	14,590	15,202	19,006	21,751
Chemical	21,604	28,479	33,838	38,896	41,560	51,591	58,976
Civil	29,932	46,972	56,246	62,803	67,293	74,322	89,718
Electrical	44,507	46,529	50,102	68,645	85,507	108,854	134,644
Mechanical	46,339	45,783	45,219	49,079	60,014	73,620	80,112
Materials	0	0	0	0	0	0	0
Other	117,136	125,745	138,663	153,445	173,695	223,378	278,302
Computer science	36,766	39,693	42,251	47,332	61,212	85,091	88,721
(Constant 1989 dollars in thousands)							
Engineering and computer science	452,706	462,272	479,095	528,556	591,630	724,134	834,876
Engineering, subtotal	398,519	408,943	425,735	471,016	519,844	627,230	736,407
Aerospace	16,031	14,602	16,455	17,736	17,828	21,644	24,141
Chemical	31,841	38,263	42,736	47,284	48,739	58,753	65,456
Civil	44,115	63,109	71,036	76,347	78,918	84,640	99,576
Electrical	65,596	62,514	63,276	83,449	100,278	123,965	149,438
Mechanical	68,296	61,511	57,109	59,663	70,381	83,840	88,915
Materials	0	0	0	0	0	0	0
Other	172,640	168,944	175,124	186,537	203,700	254,388	308,881
Computer science	54,187	53,329	53,361	57,540	71,786	96,904	98,469

TABLE 13 Continued

Field	1980	1981	1982	1983	1984	1985	1986
Federal sources							
(Current dollars in thousands)							
Engineering and computer science	679,426	766,815	812,370	873,853	951,599	1,062,822	1,210,239
Engineering, subtotal	591,956	662,620	690,927	734,914	788,679	867,105	977,578
Aerospace	42,219	43,529	49,357	53,883	54,555	61,519	72,671
Chemical	39,158	57,467	55,294	57,093	59,953	64,619	73,284
Civil	53,270	61,639	59,647	63,710	72,421	78,834	88,372
Electrical	138,639	146,763	168,357	193,315	209,652	228,549	260,340
Mechanical	94,039	94,990	97,524	100,309	118,961	134,131	148,005
Materials	0	0	0	0	0	0	0
Other	224,631	258,232	260,748	266,604	273,137	299,453	334,906
Computer science	87,470	104,195	121,443	138,939	162,920	195,717	232,661
	679,426	766,815	812,370	873,853	951,599	1,062,822	1,210,239
(Constant 1989 dollars in thousands)							
Engineering and computer science	1,001,365	1,030,250	1,025,979	1,062,306	1,115,983	1,210,366	1,343,218
Engineering, subtotal	872,448	890,259	872,603	893,404	924,920	987,479	1,084,992
Aerospace	62,224	58,483	62,335	65,503	63,979	70,059	80,656
Chemical	57,713	77,209	69,833	69,406	70,310	73,590	81,336
Civil	78,511	82,815	75,331	77,450	84,931	89,778	98,082
Electrical	204,332	197,183	212,626	235,005	245,868	260,277	288,946
Mechanical	138,598	127,623	123,167	121,941	139,511	152,751	164,267
Materials	0	0	0	0	0	0	0
Other	331,070	346,946	329,310	324,099	320,320	341,024	371,705
Computer science	128,917	139,991	153,376	168,902	191,064	222,887	258,225

Field	1987	1988	1989	1990	1991
All sources					
(Current dollars in thousands)					
Engineering and computer science	2,264,944	2,505,849	2,870,919	3,171,832	3,437,214
Engineering, subtotal	1,892,452	2,097,242	2,398,738	2,662,616	2,892,750
Aerospace	108,150	122,814	145,077	159,320	174,321
Chemical	148,362	162,559	194,060	214,887	238,553
Civil	190,873	225,265	246,509	285,113	315,134
Electrical	451,095	509,597	600,395	667,747	682,213
Mechanical	275,135	303,812	344,140	392,518	415,071
Materials	0	0	0	275,238	301,992
Other	718,837	773,195	868,557	667,793	765,466
Computer science	372,492	408,607	472,181	509,216	544,464
(Constant 1989 dollars in thousands)					
Engineering and computer science	2,436,734	2,609,172	2,870,919	3,046,323	3,427,211

Field	1987	1988	1989	1990	1991
Engineering, subtotal	2,035,989	2,183,717	2,398,738	2,557,257	2,892,747
Aerospace	116,353	127,878	145,077	153,016	174,321
Chemical	159,615	169,262	194,060	206,384	238,553
Civil	205,350	234,553	246,509	273,831	315,134
Electrical	485,309	530,609	600,395	641,324	682,212
Mechanical	296,003	316,339	344,140	376,986	415,071
Materials	0	0	0	264,347	301,992
Other	773,359	805,076	868,557	641,369	765,465
Computer science	400,744	425,455	472,181	489,066	544,463
Federal sources					
(Current dollars in thousands)					
Engineering and computer science	1,369,988	1,518,949	1,705,601	1,864,247	1,996,954
Engineering, subtotal	1,112,663	1,229,753	1,383,162	1,524,287	1,630,945
Aerospace	80,168	93,681	111,737	122,968	131,708
Chemical	76,652	85,506	101,187	107,682	114,310
Civil	89,711	103,144	101,688	116,000	122,874
Electrical	292,216	330,387	389,773	435,125	437,494
Mechanical	178,487	192,614	213,864	238,744	243,182
Materials	0	0	0	141,654	155,051
Other	395,429	424,421	464,913	362,114	426,326
Computer science	257,325	289,196	322,439	339,960	366,009
	1,369,988	1,518,949	1,705,601	1,864,247	1,996,954
(Constant 1989 dollars in thousands)					
Engineering and computer science	1,473,898	1,581,580	1,705,601	1,790,479	1,996,952
Engineering, subtotal	1,197,055	1,280,459	1,383,162	1,463,971	1,630,943
Aerospace	86,249	97,544	111,737	118,102	131,708
Chemical	82,466	89,032	101,187	103,421	114,310
Civil	96,515	107,397	101,688	111,410	122,874
Electrical	314,380	344,010	389,773	417,907	437,494
Mechanical	192,025	200,556	213,864	229,297	243,182
Materials	0	0	0	136,049	155,051
Other	425,421	441,921	464,913	347,785	426,326
Computer science	276,842	301,120	322,439	326,508	366,009
Nonfederal sources					
(Current dollars in thousands)					
Engineering and computer science	894,956	986,900	1,165,318	1,307,585	1,440,260
Engineering, subtotal	779,789	867,489	1,015,576	1,138,329	1,261,805
Aerospace	27,982	29,133	33,340	36,352	42,613
Chemical	71,710	77,053	92,873	107,205	124,243
Civil	101,162	122,121	144,821	169,113	192,260
Electrical	158,879	179,210	210,622	232,622	244,719
Mechanical	96,648	111,198	130,276	153,774	171,889
Materials	0	0	0	133,584	146,941
Other	323,408	348,774	403,644	305,679	339,140
Computer science	115,167	119,411	149,742	169,256	178,455

TABLE 13 Continued

Field	1987	1988	1989	1990	1991
(Constant 1989 dollars in thousands)					
Engineering and computer science	962,836	1,027,592	1,165,318	1,255,844	1,430,259
Engineering, subtotal	838,934	903,258	1,015,576	1,093,286	1,261,804
Aerospace	30,104	30,334	33,340	34,914	42,613
Chemical	77,149	80,230	92,873	102,963	124,243
Civil	108,835	127,156	144,821	162,421	192,260
Electrical	170,930	186,599	210,622	223,417	244,719
Mechanical	103,978	115,783	130,276	147,689	171,889
Materials	0	0	0	128,298	146,941
Other	347,938	363,155	403,644	293,583	339,140
Computer science	123,902	124,335	149,742	162,559	178,455

NOTE: Before 1980, NSF did not collect data by field of engineering.
SOURCE: NSF CASPAR Database System.

TABLE 14 Full-Time Graduate Research Assistants by Field and Source of Support, 1972-1991

Field	1972	1973	1974	1975	1976	1977	1978	1979
All sources of support								
Engineering and computer science	10,369	11,033	11,850	11,733	12,059	12,543	0	13,634
Engineering, subtotal	9,731	10,380	11,103	10,987	11,328	11,819	0	12,817
Aerospace	598	544	534	512	484	497	0	503
Chemical	1,326	1,351	1,348	1,364	1,455	1,487	0	1,745
Civil	1,391	1,607	1,836	1,718	1,858	1,957	0	1,926
Electrical	2,093	2,203	2,187	2,183	2,153	2,435	0	2,596
Mechanical	1,200	1,556	1,696	1,636	1,778	1,762	0	1,967
Materials	1,000	987	1,018	1,070	1,105	1,150	0	1,320
Industrial	467	494	616	516	476	460	0	563
Other	1,656	1,638	1,868	1,988	2,019	2,071	0	2,197
Computer science	638	653	747	746	731	724	0	817
Federal sources, total								
Engineering and computer science	6,821	7,005	7,258	7,387	7,599	7,971	0	8,579
Engineering, subtotal	6,416	6,554	6,781	6,935	7,213	7,497	0	7,998
Aerospace	414	407	406	385	330	366	0	355
Chemical	777	779	707	817	863	899	0	950
Civil	789	855	962	902	1,019	1,025	0	1,094
Electrical	1,638	1,668	1,646	1,697	1,665	1,828	0	1,878
Mechanical	743	979	1,010	1,045	1,185	1,189	0	1,273
Materials	755	756	740	760	823	836	0	1,013
Industrial	260	198	266	220	172	212	0	230
Other	1,040	912	1,044	1,109	1,156	1,142	0	1,205
Computer science	405	451	477	452	386	474	0	581

TABLE 14 Continued

Field	1972	1973	1974	1975	1976	1977	1978	1979
National Science Foundation								
Engineering and computer science	2,207	2,390	2,325	2,364	2,284	2,314	0	2,368
Engineering, subtotal	2,002	2,174	2,134	2,173	2,138	2,094	0	2,128
Aerospace	58	60	59	75	48	45	0	21
Chemical	369	437	376	395	404	375	0	345
Civil	246	241	229	251	279	262	0	316
Electrical	494	549	535	556	513	493	0	552
Mechanical	217	319	344	320	311	329	0	300
Materials	245	243	239	274	274	291	0	351
Industrial	52	64	69	57	48	45	0	39
Other	321	261	283	245	261	254	0	204
Computer science	205	216	191	191	146	220	0	240
National Institutes of Health								
Engineering and computer science	381	334	350	424	472	387	0	367
Engineering, subtotal	356	299	314	385	430	348	0	327
Aerospace	3	7	6	5	6	4	0	1
Chemical	50	40	36	32	50	40	0	40
Civil	7	10	6	7	2	11	0	4
Electrical	118	75	106	111	125	111	0	93
Mechanical	49	50	34	60	66	45	0	52
Materials	31	37	28	32	36	24	0	27
Industrial	39	6	15	18	6	7	0	10
Other	59	74	83	120	139	106	0	100
Computer science	25	35	36	39	42	39	0	40

Other HHS								
Engineering and computer science	102	97	116	109	73	73	0	106
Engineering, subtotal	101	93	108	98	66	66	0	105
Aerospace	1	0	0	2	0	0	0	2
Chemical	14	15	9	8	4	3	0	9
Civil	7	16	14	5	19	9	0	22
Electrical	25	4	16	18	14	7	0	13
Mechanical	27	29	19	18	12	21	0	6
Materials	6	2	12	10	2	5	0	10
Industrial	6	14	22	22	6	5	0	17
Other	15	13	16	15	9	16	0	26
Computer science	1	4	8	11	7	7	0	1
Department of Defense								
Engineering and computer science	1,726	1,558	1,650	1,627	1,632	1,705	0	1,777
Engineering, subtotal	1,609	1,424	1,464	1,470	1,474	1,538	0	1,569
Aerospace	185	143	160	138	133	141	0	154
Chemical	66	44	51	49	33	36	0	47
Civil	67	46	58	72	64	58	0	46
Electrical	622	628	574	641	635	712	0	723
Mechanical	145	201	180	186	208	218	0	205
Materials	181	183	158	165	186	174	0	182
Industrial	92	46	51	48	31	47	0	47
Other	251	133	232	171	184	152	0	165
Computer science	117	134	186	157	158	167	0	208

TABLE 14 Continued

Field	1972	1973	1974	1975	1976	1977	1978	1979
Other federal agencies								
Engineering and computer science	2,405	2,626	2,817	2,863	3,138	3,492	0	3,961
Engineering, subtotal	2,348	2,564	2,761	2,809	3,105	3,451	0	3,869
Aerospace	167	197	181	165	143	176	0	177
Chemical	278	243	235	333	372	445	0	509
Civil	462	542	655	567	655	685	0	706
Electrical	379	412	415	371	378	505	0	497
Mechanical	305	380	433	461	588	576	0	710
Materials	292	291	303	279	325	342	0	443
Industrial	71	68	109	75	81	108	0	117
Other	394	431	430	558	563	614	0	710
Computer science	57	62	56	54	33	41	0	92
Nonfederal sources								
Engineering and computer science	3,548	4,028	4,592	4,346	4,460	4,572	0	5,055
Engineering, subtotal	3,315	3,826	4,322	4,052	4,115	4,322	0	4,819
Aerospace	184	137	128	127	154	131	0	148
Chemical	549	572	641	547	592	588	0	795
Civil	602	752	874	816	839	932	0	832
Electrical	455	535	541	486	488	607	0	718
Mechanical	457	577	686	591	593	573	0	694
Materials	245	231	278	310	282	314	0	307
Industrial	207	296	350	296	304	248	0	333
Other	616	726	824	879	863	929	0	992
Computer science	233	202	270	294	345	250	0	236

Field	1980	1981	1982	1983	1984	1985	1986	1987	1988	1989	1990	1991
All sources of support												
Engineering and computer science	14,959	15,486	15,799	16,940	17,919	19,976	22,766	24,984	26,499	27,941	28,425	30,311
Engineering, subtotal	13,923	14,388	14,608	15,537	16,284	17,900	20,412	22,147	23,452	24,602	25,086	26,763
Aerospace	580	583	617	691	673	725	823	815	934	1,040	1,137	1,232
Chemical	1,949	2,136	2,199	2,413	2,487	2,605	2,741	2,970	3,007	3,026	3,017	3,180
Civil	2,121	2,111	2,027	2,245	2,440	2,417	2,786	2,908	3,072	3,042	3,095	3,562
Electrical	2,851	2,891	2,950	3,192	3,156	3,677	4,447	5,111	5,722	6,129	6,212	6,576
Mechanical	2,052	2,138	2,213	2,371	2,663	3,280	3,666	3,930	4,069	4,248	4,238	4,633
Materials	1,390	1,558	1,522	1,681	1,749	1,963	2,247	2,264	2,331	2,507	2,545	2,509
Industrial	591	542	552	433	563	585	716	944	1,049	1,167	1,130	1,270
Other	2,389	2,429	2,528	2,511	2,553	2,648	2,986	3,205	3,268	3,443	3,712	3,801
Computer science	1,036	1,098	1,191	1,403	1,635	2,076	2,354	2,837	3,047	3,339	3,339	3,548
Federal sources, total												
Engineering and computer science	9,212	9,240	9,325	9,838	9,646	9,499	10,704	11,868	12,617	13,042	12,941	13,865
Engineering, subtotal	8,534	8,525	8,562	8,995	8,675	8,426	9,556	10,361	10,971	11,259	11,153	11,910
Aerospace	375	399	469	541	524	478	558	567	634	623	651	741
Chemical	1,100	1,136	1,121	1,213	1,196	1,141	1,245	1,366	1,325	1,267	1,287	1,310
Civil	1,237	1,076	1,059	1,058	1,109	1,009	1,117	1,225	1,221	1,198	1,176	1,406
Electrical	1,982	1,971	2,061	2,124	1,829	1,694	2,027	2,332	2,750	2,752	2,719	2,933
Mechanical	1,240	1,283	1,278	1,423	1,469	1,585	1,709	1,926	2,061	2,136	2,044	2,158
Materials	1,017	1,132	1,067	1,139	1,111	1,146	1,323	1,240	1,288	1,411	1,327	1,235
Industrial	261	240	189	167	169	156	178	248	289	289	271	375
Other	1,322	1,288	1,318	1,330	1,268	1,217	1,399	1,457	1,403	1,583	1,678	1,752
Computer science	678	715	763	843	971	1,073	1,148	1,507	1,646	1,783	1,788	1,955

TABLE 14 Continued

Field	1980	1981	1982	1983	1984	1985	1986	1987	1988	1989	1990	1991
National Science Foundation												
Engineering and computer science	2,447	2,465	2,626	2,749	2,753	2,773	3,183	3,626	3,942	3,954	3,946	4,287
Engineering, subtotal	2,171	2,133	2,285	2,411	2,370	2,355	2,811	3,119	3,405	3,289	3,245	3,508
Aerospace	31	27	32	36	24	24	30	55	62	55	73	75
Chemical	355	421	452	509	518	512	552	587	590	550	542	534
Civil	326	310	373	361	357	336	377	442	390	349	392	443
Electrical	532	464	541	546	459	441	635	736	1,003	929	936	1,030
Mechanical	263	270	284	323	355	390	435	538	608	559	528	594
Materials	342	361	329	367	342	385	410	402	418	431	373	330
Industrial	35	41	30	40	54	38	76	94	110	118	116	164
Other	287	239	244	229	261	229	296	265	224	298	285	338
Computer science	276	332	341	338	383	418	372	507	537	665	701	779
National Institutes of Health												
Engineering and computer science	373	325	347	368	374	347	377	427	515	578	600	628
Engineering, subtotal	333	293	311	352	354	327	335	377	459	537	558	585
Aerospace	1	4	3	2	4	1	2	5	2	3	9	14
Chemical	30	34	35	37	56	44	60	58	78	75	76	72
Civil	19	8	17	16	7	8	16	7	3	13	9	29
Electrical	89	75	50	59	63	54	49	46	93	70	82	85
Mechanical	37	33	44	51	42	59	38	51	55	53	62	49
Materials	22	14	7	16	16	17	10	5	7	9	13	12
Industrial	8	5	10	10	9	8	4	6	5	16	21	15
Other	127	120	145	161	157	136	156	199	216	298	286	309
Computer science	40	32	36	16	20	20	42	50	56	41	42	43

Other HHS												
Engineering and computer science	57	45	55	59	44	44	42	83	56	61	70	49
Engineering, subtotal	54	43	51	56	43	43	41	83	56	55	61	43
Aerospace	0	0	0	0	0	1	0	4	3	0	0	0
Chemical	5	1	8	5	7	1	2	8	4	0	2	2
Civil	21	20	6	10	0	9	2	4	2	2	3	4
Electrical	8	6	22	20	12	16	16	24	24	17	21	10
Mechanical	6	8	6	2	4	14	6	11	2	4	1	2
Materials	1	1	0	1	3	0	0	2	2	0	1	4
Industrial	11	6	4	1	0	0	8	5	7	12	4	13
Other	2	1	5	17	17	2	7	25	12	20	29	8
Computer science	3	2	4	3	1	1	1	0	0	6	9	6
Department of Defense												
Engineering and computer science	1,895	2,189	2,365	2,642	2,574	2,633	3,052	3,811	3,924	3,927	3,644	3,760
Engineering, subtotal	1,686	1,903	2,082	2,325	2,146	2,169	2,523	3,053	3,104	3,127	2,936	3,007
Aerospace	161	221	197	241	234	231	250	294	306	276	268	292
Chemical	64	50	66	87	83	79	88	132	123	111	79	113
Civil	54	44	60	74	97	104	149	189	192	194	153	175
Electrical	749	865	953	1,034	862	822	906	1,116	1,141	1,150	1,110	1,106
Mechanical	231	261	262	362	344	381	428	586	612	658	546	552
Materials	191	206	251	249	278	305	420	409	380	408	425	412
Industrial	60	58	48	42	30	44	36	56	65	52	49	50
Other	176	198	245	236	218	203	246	271	285	278	306	307
Computer science	209	286	283	317	428	464	529	758	820	800	708	753

TABLE 14 Continued

Field	1980	1981	1982	1983	1984	1985	1986	1987	1988	1989	1990	1991
Other federal agencies												
Engineering and computer science	4,440	4,216	3,932	4,020	3,901	3,702	4,050	3,921	4,180	4,522	4,681	5,141
Engineering, subtotal	4,290	4,153	3,833	3,851	3,762	3,532	3,846	3,729	3,947	4,251	4,353	4,767
Aerospace	182	147	237	262	262	221	276	209	261	289	301	360
Chemical	646	630	560	575	532	505	543	581	530	531	588	589
Civil	817	694	603	597	648	552	573	583	634	640	619	755
Electrical	604	561	495	465	433	361	421	410	489	586	570	702
Mechanical	703	711	682	685	724	741	802	740	784	862	907	961
Materials	461	550	480	506	472	439	483	422	481	563	515	477
Industrial	147	130	97	74	76	66	54	87	102	91	81	133
Other	730	730	679	687	615	647	694	697	666	689	772	790
Computer science	150	63	99	169	139	170	204	192	233	271	328	374
Nonfederal sources												
Engineering and computer science	5,747	6,246	6,474	7,102	8,273	10,477	12,062	13,116	13,882	14,899	15,484	16,446
Engineering, subtotal	5,389	5,863	6,046	6,542	7,609	9,474	10,856	11,786	12,481	13,343	13,933	14,853
Aerospace	205	184	148	150	149	247	265	248	300	417	486	491
Chemical	849	1,000	1,078	1,200	1,291	1,464	1,496	1,604	1,682	1,759	1,730	1,870
Civil	884	1,035	968	1,187	1,331	1,408	1,669	1,683	1,851	1,844	1,919	2,156
Electrical	869	920	889	1,068	1,327	1,983	2,420	2,779	2,972	3,377	3,493	3,643
Mechanical	812	855	935	948	1,194	1,695	1,957	2,004	2,008	2,112	2,194	2,475
Materials	373	426	455	542	638	817	924	1,024	1,043	1,096	1,218	1,274
Industrial	330	302	363	266	394	429	538	696	760	878	859	895
Other	1,067	1,141	1,210	1,181	1,285	1,431	1,587	1,748	1,865	1,860	2,034	2,049
Computer science	358	383	428	560	664	1,003	1,206	1,330	1,401	1,556	1,551	1,593

NOTE: Data were not collected in 1978.
SOURCE: NSF CASPAR Database System.

Biographical Information

COMMITTEE MEMBERS

WILLIAM SCHOWALTER (Chairman) is dean of the College of Engineering at the University of Illinois at Urbana-Champaign. He earned a B.S. from the University of Wisconsin and an M.S. and a Ph.D from the University of Illinois. He has been a member of the National Academy of Engineering since 1982 and has participated in a number of NAE activities, including the Panel on Engineering Research Centers, the Academic Advisory Board, Awards Committee, and the Committee on Membership.

DANIEL C. DRUCKER is emeritus professor of aerospace engineering, mechanics, and engineering science at the University of Florida at Gainesville. Before joining the University of Florida, he was dean of the College of Engineering at the University of Illinois at Urbana-Champaign from 1968 to 1984. Prior to this, Dr. Drucker served on the engineering faculty at Brown University for 21 years. He earned his B.S., C.E., and Ph.D. degrees from Columbia University. Dr. Drucker has been a member of the National Academy of Engineering since 1967.

ALEXANDER FLAX is a senior fellow at the National Academy of Engineering. From 1963 to 1969, he was assistant secretary of the Air Force for R&D, and from 1965 to 1969, he also held the post of director, National Reconnaissance Office. In 1969, Dr. Flax joined the Institute for Defense Analyses and became its president that same year, serving in that position until 1983. He earned a bachelor's degree from the Guggenheim School of

Aeronautics of New York University, and a Ph.D. from the University of Buffalo. Dr. Flax has been a member of the National Academy of Engineering since 1967 and served as the NAE Home Secretary from 1984 to 1992.

WILLIAM C. GEAR is president of NEC Research Institute in Princeton, New Jersey. Dr. Gear began his career as an engineer at IBM British Laboratories in Hurselyat. He was a professor of computer science and applied mathematics at the University of Illinois at Urbana-Champaign from 1962 to 1985. In 1985, he was named head of the university's computer science department. He earned an M.S. and a Ph.D. from the University of Illinois at Urbana-Champaign and an M.A. and a B.A. from Cambridge University in England. He has been a member of the National Academy of Engineering since 1992.

PAUL C. JENNINGS is vice president and provost as well as professor of civil engineering and applied mechanics at the California Institute of Technology. Prior to becoming vice president and provost in 1989, Dr. Jennings was the chairman of the division of engineering and applied science at Cal Tech and has been a professor at that institution since 1966. He earned a B.S. from Colorado State University and an M.S. and a Ph.D. from Cal Tech. Dr. Jennings has been a member of the National Academy of Engineering since 1977.

RICHARD SEEBASS is professor of aerospace engineering sciences at the University of Colorado in Boulder. From 1981 to 1994, he was the dean of the College of Engineering and Applied Science at the University of Colorado. From 1958 to 1977, Dr. Seebass was a professor of mechanical and aerospace engineering at Cornell University. He earned a B.S.E. and an M.S.E. from Princeton University and a Ph.D. from Cornell University. He has been a member of the National Academy of Engineering since 1985.

JOHN A. WHITE is dean of the College of Engineering at Georgia Tech and has been a member of the Georgia Tech faculty since 1975. From 1988 to 1991, he served as assistant director for engineering at the National Science Foundation. Dr. White earned a Ph.D. from Ohio State University, an M.S.I.E. from Virginia Polytechnic Institute, and a B.S.I.E. from the University of Arkansas. He has been a member of the National Academy of Engineering since 1987.

SPEAKERS

DUANE A. ADAMS serves as the deputy director of the Advanced Research Projects Agency (ARPA). Before coming to ARPA in 1992, Dr. Adams was the associate dean of research at the School of Computer Science at Carnegie Mellon University (CMU). Prior to joining the CMU faculty, Dr. Adams served for 20 years in the Air Force. He earned a B.A. from the University of Montana, an M.A. from the University of California, Berkeley, and a Ph.D. from Stanford University.

JOHN A. ARMSTRONG, retired vice president for science and technology at IBM Corporation, was the 1993–94 Karl Taylor Compton Lecturer at the Massachusetts Institute of Technology and is visiting professor of electrical engineering and computer science at the University of Virginia in Charlottesville. He received A.B. and Ph.D. degrees from Harvard College and joined IBM in 1963 as a member of the firm's research staff. In 1986, Dr. Armstrong was named IBM's director of research. He became an IBM vice president in 1987 and was elected a member of the Corporate Management Board in 1989. He retired from IBM in 1993. Dr. Armstrong is a member of the National Academy of Engineering and of the Royal Swedish Academy of Engineering Sciences.

NEAL F. LANE began his 6-year term as director of the National Science Foundation (NSF) in October 1993. Before assuming that position, Dr. Lane was provost and professor of physics at Rice University, a position he had held since 1986. His tenure at Rice began in 1966, when he joined the department of physics as an assistant professor. Dr. Lane has also served briefly as chancellor of the University of Colorado at Colorado Springs and as director of the division of physics at the NSF. He is widely recognized as a scientist and an educator, having served as president of Sigma Xi and twice receiving Rice University's George R. Brown Prize for Superior Teaching. Dr. Lane holds B.S., M.S., and Ph.D. degrees from the University of Oklahoma.

SIMON OSTRACH serves as the home secretary of the National Academy of Engineering and since 1970 has been the Wilbert J. Austin Distinguished Professor of Engineering at Case Western Reserve University. From 1950 to 1960, Dr. Ostrach was the chief of the fluid physics branch at the National Aeronautics and Space Administration. Dr. Ostrach earned a B.S. and an M.E. from the University of Rhode Island and a Sc.M. and a Ph.D. from Brown University. Dr. Ostrach received an honorary D.Sc. Technion from Israel Institute of Technology and an honorary D.Eng. from Florida State University. Dr. Ostrach has been a member of the National Academy of Engineering since 1978.

CHANG-LIN TIEN is chancellor and A. Martin Berlin Professor at the University of California, Berkeley. Dr. Tien joined the Berkeley faculty in 1959 as an acting assistant professor of mechanical engineering. He later became a full professor and chairman of the department and was for 2 years Berkeley's vice chancellor for research. In 1990, he became Berkeley's seventh chancellor, the first Asian-American to head a major U.S. research university. In 1962, at age 26, he became the youngest professor to win Berkeley's Distinguished Teaching Award. He earned an M.A. at the University of Louisville and an M.A. and a Ph.D. at Princeton University. He has been a member of the National Academy of Engineering since 1976.